MW01378504

Copyright © 2019 Upstate Venture Connect

All rights reserved.

Published in the United States by UVC Press.

All trademarks are the property of their respective companies.
All photographs are with permission.

ISBN: 978-1-7331448-0-3

PROUDLY PRINTED IN THE USA

For Upstate's emerging entrepreneurs ~
may these stories light the path to your success.

OneGroup is excited to be the lead sponsor of the first Upstate Founder's Playbook. We are inspired by the hardworking innovators featured in the pages ahead and honored to be associated with like-minded growth companies that are committed to Upstate New York. Our success helping technology driven firms such as Saab-Sensis and Digital Hyve plus our involvement with technology incubator programs, made our decision to support Upstate Venture Connect easy.

The companies featured in this limited-edition book are from a wide range of industries yet they share a common goal: bringing to market innovative solutions at scale through technology. They're Upstate-grown and attracting the best and brightest minds to our region.

OneGroup itself is a leader in our industry and one of the fastest growing risk management and insurance firms in Upstate New York, now with 18 locations. Like the outstanding companies featured within, OneGroup's unique business model of assembling a wide range of expertise to collaborate on solving client issues, also requires us to continually increase our talent pool in Upstate New York. Our groundbreaking, service-based approach to business solutions focuses on the *true* needs of those we serve – not perceived needs, price or product.

As a reader of this publication, we thank you for supporting the growth of emerging companies in Upstate New York. Their unique stories provide insight into their drive to succeed, their challenges and how they are managing their growth.

OneGroup is proud to assist in bringing you this book and making it possible to learn about these entrepreneurs. We wish each of them outstanding success in their quest to help others through their unique offerings.

Pierre Morrisseau, CEO

Table of Contents

Foreword

We created this Playbook because the stories need to be told.

Rapid de-industrialization left Upstate New York communities with a rich and storied history, but no clear path to the future. Kodak and Xerox have given way to Instagram and Snapchat while once mighty IBM and GE struggle for relevance in the era of pocket supercomputers and alternative energy.

The founders interviewed for this playbook are showing us the way forward. Their globally competitive companies are based in Upstate, creating high-paying jobs and taking advantage of a great quality of life. Their stories are in sharp contrast to those who believe the deck is stacked against the region due to some cruelty of nature or misgovernment from Albany.

This inaugural publication showcases founders who reflect Upstate's enormous creative capacity and diverse opportunities for building fast scaling enterprises. They are also in the vanguard of a much larger cohort of innovative companies sprouting across the region.

The founders' journeys we discovered are unique, as are the obstacles they encountered. As we reflected upon these conversations, however, a set of shared themes emerged surrounding their context, choices and mindset.

Adapting to Upstate Context

Capital Access—Availability of growth capital is often cited as a barrier to new venture formation in Upstate. We could always use more local venture capital funds, but that will only be possible when we have a lot more fundable startups. The founders we spoke to span the entire spectrum of capital intensity: AIS, SpinCar, Ephesus and Vicarious Visions raised minimal amounts of funding to get to profitability; ACV Auctions, Athenex, Kionix and Rheonix needed hundreds of millions of dollars to scale. Importantly, each of these companies was able to find the money to scale once they had local support to get off the ground.

Distant Customers—Another shared context for each of these founders is that their customers were spread across the world (Guitar Hero players) or concentrated in far away places (like CVC customers in Silicon Valley). Being on the road for these founders was the norm, and at every point they were able to compete and overturn the status quo.

Distributed Teams—Having a world-class team was consistently identified as critical to the success of each venture. In the case of capital intensive solutions with long lead times to market (e.g., Rheonix, Athenex), the startup team has to have the reputation and capacity to raise tens (or even hundreds) of millions of dollars without having a product ready for sale. For others, product ingenuity needs to be complemented with a high performing sales function. These leaders put the needs of talent first and built teams that were often distributed across the country and (in the case of Athenex) around the world.

Founder Choices

Investor Equity vs Execution Speed—Many of the founders interviewed had to raise substantial sums of capital. They faced the choice of diluting their equity interests in order to acquire cash to fuel their plans. In nearly every case, they realized that the speed with which their company grew in value outweighed the potential dilution. The decision was clear: If you have the right investor partner, focus on driving growth and the rest takes care of itself.

Exit Options—The range of exits for each of these firms also spans the spectrum from IPO (Synacore, CVC, Athenex) to strategic acquisitions (SpinCar, iCardiac, Vicarious Visions, Ephesus). Typically, the right option emerged when the founders were ready to consider a change of control in return for more growth capital access and/or the possibility of creating some liquidity for their investors.

Entrepreneur Mindset

Global vs Local—Each of the founders featured here is deeply rooted in their local community, but has a global view of their business. They are all clear-eyed about the challenges of living in communities with few direct flights and scarce growth capital, but are also aware that life for an emerging tech company is not a cakewalk in any location. By focusing their efforts on being better than all of their competitors, these founders effectively made location a non-issue for prospective employees, customers, investors and acquirers.

Relentless Optimism + Tireless Effort—Building something new is fraught with technical challenges and convincing customers to change their ways can be exhausting. Successful founders are infected with a strain of positive thinking that wills them to succeed against all odds. They also share a fanatical commitment to understanding their customers and serving them better than the competition.

Hooked on Building—Another common trait across founders is their in-built attraction to new ideas and opportunities. All of them could have had successful careers in any organization, but they gravitated towards being on the ground floor of launching new ideas. Setbacks were part of the game and never a reason to stop. Success in one venture meant it was time to start the next adventure.

Paying it Forward—Each of the founders interviewed exhibited a deep sense of gratitude to the numerous people who helped and guided them on their way. Not surprisingly, all of them are paying it forward by helping other founders as mentors, investors, and cheerleaders.

Since 2010, Upstate Venture Connect has focused on accelerating Upstate NY founders' access to the right people and resources. We have learned that Upstate's best asset is also its most mobile—namely, a half-million students being educated at more than 100 colleges and universities. Every company referenced here is a direct beneficiary of the research and talent that sets our region apart. And every founder firmly believes in supporting the ambitions of future founders. That is a simple formula for building a growing entrepreneurial economy. We hope these stories will inspire readers to start their own world beating companies or also help others on their path to success.

This Playbook was made possible by an army of supporters who share our vision of a brighter and more prosperous region. We are grateful that the founders made the time to share and reflect upon their experiences, we want to thank OneGroup leaders for believing in the value of this project. Our colleagues Martin Babinec, Kathryn Cartini and Marylou Herringshaw also deserve special recognition because of their constant commitment to the UVC cause. Most of all, we owe a huge debt to our families and the extraordinary community of Upstate New York startup supporters who encourage and inspire us every day.

We have so much to accomplish together!

Jennifer Sertl and Nasir Ali
Rochester, NY
June 2019

We believe in giving feedback directly and as often as possible.

Guha Bala

Just to hear someone say "You can do it," removes that mental barrier.

Karthik Bala

Guha Bala & Karthik Bala

Guha and Karthik Bala grew up in Brighton just outside Rochester, NY. In 1991, the two brothers founded Vicarious Visions, a revered developer of iconic video games like Guitar Hero. Vicarious Visions was acquired by Activision Publishing in 2005. Guha and Karthik stayed on to grow the company in the Albany area and generate billions in revenues over the next decade while also laying the foundation of a nationally recognized game development cluster.

In 2017-18, the Bala brothers left Activision to form their own investment firm Velan Ventures and game development shop Velan Studios in Troy, NY. Guha graduated from Harvard University with an AB in Chemistry. Karthik graduated from Rensselaer Polytechnic Institute with a Bachelor's degree in Computer Science and Psychology. Both brothers hold MBAs from the MIT School of Management.

Sertl: *Okay. Let's begin with Vicarious. What was your vision for the business and why did you start the company?*

K. Bala: Vicarious Visions was started in our parents' basement in Rochester. I was 15 and Guha was 14 when we started the company because we just wanted to make games. We loved playing games and making things. It was 1991 and we did not even have internet access at the time. Knowledge and information about making games was limited. At the time, it was not about building a business. We were just passionate about making a game. It started as a hobby then took on a life of its own.

Karthik Bala and the team circa 1998 at the RPI Incubator.

Sertl: *And how did the vision shift over time?*

K. Bala: The vision really started around a specific game that we wanted to make, and a story that we wanted to tell within that game. It was a product-based vision. It was not a company vision. Guha and I had a few friends and some freelance contractors working with us. It was all about getting that first game made. We were able to secure a publishing deal for that game while I was a freshman at Rensselaer Polytechnic Institute

(RPI) and Guha was a high school senior. That gave us some level of financing to be able to finish the product. After five years, we had the initial idea complete and our product shipped. We never got the royalties that we were owed and were left with a lot of debt. We had to decide what to do next. That was an important moment because that is when we realized that it was not just about making a good product. We were proud of what we had made but it got a mixed reception both critically and commercially. We needed to make great products that were also successful in the marketplace. It was then that we realized we had to build a company.

Global games that got their start in Upstate New York with Vicarious Visions.

Sertl: *How would you define a creative enterprise versus a corporate enterprise?*

G. Bala: We were doing the first game as artists – for art's sake. We thought that whatever we made, people would buy. We thought that because we really liked it, there would be an audience for it. We had a unique vision for the game and thought because of the creativity, of course people would want to buy it. That was not the case. Fewer people loved it than we thought. It took us a while to figure it out, but we decided that we had to do games on a platform to reach larger audiences. We would have to speak to universal fantasies in narratives. We had to create a vision that was beyond our personal satisfaction and expression that could be shared and enjoyed by many. Bringing our creative vision in line with what we felt could be much more resonant with a bigger audience was a bit of a journey. A product for a potentially large audience that would resonate and be a hit was going to require enough economics in it for us to keep reinvesting. We had to bring in the right talent around us because our capabilities could only go so far. If we wanted to do things that our customers wanted, we needed to hire people that had the same or better capabilities in almost everything that we did. The commercial enterprise was in service of building a great creative enterprise.

Sertl*: You could have built a company anywhere. Why did you choose to build in the capital region?*

K. Bala: We met Mike Marvin, who at the time was Chairman of MapInfo Corporation. MapInfo was a great success story coming out of Rensselaer Polytechnic Institute, in the Capital Region. MapInfo had gone public. Mike was an active angel investor and mentor in the community supporting tech startups. We met him when he was part of an angel fund called Exponential. After our first game was published, we were in debt and trying to figure out what to do next. We told him that we wanted to keep things going and build a real company. We realized we needed funding but even more importantly we needed mentoring, support and guidance to build a company and understand what that means. We were fortunate that Marvin agreed to help us under the condition that we stay here in the Capital Region to help build the tech ecosystem in Upstate New York. That was the handshake deal. Subsequent to Marvin coming on our board, he was one of our mentors and we were able to get some angel funding through his group.

Sertl: *Now that you are such a big part of our ecosystem, what are the benefits of building in Upstate New York?*

K. Bala: There are a lot of interesting pieces here. Quality of life is certainly a key component. Having grown a business and having a family with kids makes you appreciate it even more. The quality of life aspect is great. There is also a tremendous amount of university talent for startups. Those two aspects are important. In our industry, we have been growing a critical mass and building a cluster for video games here in the Capital Region. Critical mass is an important factor for growth.

G. Bala: Whether it is Albany, Syracuse, Rochester or Buffalo, these communities tend to be inclusive. They tend to be a bit more cosmopolitan as well as open. Encouraging people from all walks of life from various backgrounds is important for creative businesses to flourish. We need an environment that facilitates creative collisions. We have a supportive business ecosystem and people genuinely want to help other businesses prosper and thrive here. That is what we found when we started. We did not have much, but we did have a business community that really wanted us to succeed and went out of their way to open their doors. I think for game companies there is a more established ecosystem in Upstate New York today between RPI and local game companies that were not here when we started. I think that building high quality engineering teams in the middle of a large metro area is exceedingly difficult if you are a startup now because of the intensity of competition from the big tech companies such as Google and Facebook. With heavily funded venture backed companies, the salary escalation is dramatic. For example, Warner Brothers, Inc. has twenty full stack network engineers across the hall from us at Velan Studios, Inc. It would be hard for them to build that anywhere else in the country. We could do it here because there is no immediate competition from big tech. It is an interesting ecosystem advantage.

Sertl: *What are the challenges of building in Upstate New York?*

G. Bala: What we have here is an opportunity to get to a much bigger scale. It is not something we can take a lot of comfort in. All of these quality of life aspects are meaningful to me now because I am in my 40s. For a person that is 20, 25 or even up to 35 years old, it doesn't matter that you can have affordable housing. It doesn't matter that you have a car. You don't need to have a big place to live and that kind of thing. The younger end of our demographic values density of amenities and cultural density. The urban centers are spread and Rochester is a few hours away from Albany. In a place like Union Square in New York City or in Cambridge, Austin or any number of metros, there is a density of startup companies and cultural opportunities that resonate with the key demographic for the talent we need. We have to work harder to get that talent here. The other thing is that even though in the metro areas there is a high cost to having a startup, if your idea does not work you have a better chance to exit quickly, refactor and keep going. People reinvent at a much more rapid rate in high cost locations like New York City and Silicon Valley. Here, people tend to keep with their ideas whether they are working or not.

Sertl: *What are lessons learned from your funding path?*

K. Bala: One of the key things is to do something distinctive and not just be chasing trends. If you try to chase a trend you are always going to be behind. Focus on key talent and work on something distinctive and that will help you cut through the noise. You have got to be really focused, especially in early seed capital or Series A stages. You can only differentiate if you are focused. If you are differentiated and focused and have a commercial breakthrough, it is possible to access capital to help you grow.

G. Bala: It is a bit easier to get funding based on trends and general segment bits where there is already a lot of venture activity, for example in the centers of investment in much larger metros. Here, we have to be more disciplined in fundamentals. We have to be clear on the value proposition, the business model, and the gross margins. It is important to be able to prove that you can stay laser focused on fundamentals. I still see a lot of business plans based on the trends that use numbers in the aggregate, large market sizes, et cetera, and that focus less on the fundamentals. We have always grown our businesses on fundamentals, and this has made us economically resilient. The fundamentals enabled us to be attractive for funding and allowed us to pick the right kind of partners. Building a solid fundamentals strategy is something that is much more suitable for locations like ours.

Sertl: *From 1991 through 2005, how many employees did you ultimately have?*

K. Bala: We had about ninety employees when we sold the business to Activision.

Vicarious Visions Team

Sertl: *How did you know it was time to sell?*

G. Bala: Our sale happened at the beginning of 2005. We had always run our business without any intention to sell. It was all about building a great business. In 2003, we began noticing some significant shifts in the industry. To make great games, the teams had to be bigger and able to produce more intense multimedia experiences. We wanted to be at the high end of the quality range because in the United States, you cannot compete with a low-cost strategy. This translated to fewer partners and bigger projects. Many of our customers were consolidating because there were fewer and bigger bets being made. And at that time the names were Electronic Arts, Activision, THQ, Ubisoft. Activision was one of our customers. We started doing business with them in 1999.

Sertl: *How did you decide to sell to Activision?*

G. Bala: The sale of the business was more of a strategic progress around the vision of where our company was going. The most strategic piece was thinking about a partner that would be suitable. There were two other things we had to consider: financial and cultural fit. It turned out that our economics were improving and phenomenal on a year-on-year basis. We had made several multi-platinum hits over the five years that we had been working with Activision. We had remarkable cultural compatibility including their approach to business. Their approach was an independent studio model which allowed creative companies to operate with autonomy. They were focused on getting better and better. They also prioritized quality. We got along with the CEO, Chairman and all the way down to the line people within Activision. We were aligned on strategy, finance, and culture.

Sertl: *Is there anything that you can teach us about how to build a vision and engage others in that vision?*

K. Bala: You have to believe the vision yourself. You have to be passionate. It is more than articulating a vision. You have to share the vision and make it crisp. You have to keep iterating. You have to feel that it is right in your bones and you have to pressure test it. Find people who feel the same way and build that critical mass and momentum towards realizing it. A key part of it is a philosophy of surrounding yourself with people that are better than you. As the leader, you have to listen well. You have to listen to as many experts as you can.

G. Bala: I have a couple of notions about vision. Going back to something Karthik said earlier, to have a distinctive view of the world – it has to mean something. The big picture has to say something unique, distinct and compelling. The way we evolved our vision is by coming up with something deeply meaningful to us. It was a very active choice. It is difficult to achieve because other people have not done it yet. Then we had to figure out how can that vision resonate with the people around us? What are the elements of the vision that can get resonance with all the people that are going to buy into it? We go through a process for being able to refine and say, "Does this even work for anybody?" We had to find the right words and put them against the plan. It is important for a vision to say, "here is where we are going, here are the steps we are taking to get there, and here's how our people are key to getting there." This also ensures their stake in the outcome. People need to see themselves in that vision and see themselves on that journey.

Sertl: *What have been some unexpected situations that you had to overcome? How did you build resilience in your culture?*

K. Bala: That happens about once a quarter. In 2000, we had around 45 employees and were building momentum. We built a management team as we realized it was not just the two of us anymore. We had to build real structure with somebody heading up development, human resources and all the important aspects of the company. After we had made that investment in our infrastructure, our biggest customer went bankrupt. Overnight, ninety percent of our revenue disappeared. That was a wake-up call. We were in a lot of trouble because of our huge payroll. Guha and I were three years out of college and we had to get a million-dollar personal bank loan. We did not miss a single payroll with the support of M&T Bank. At the same time we had to keep the team focused because quality mattered more than ever. We had to replace that lost business with a new set of customers. Fortunately, we paid off that loan the following year when we had our first multi-million unit, critically acclaimed game, Tony Hawk's Pro Skater 2 for the Game Boy Advance. It was the first of many, many scary times.

Sertl: *You talked about the team. How did your people management process change as you grew?*

G. Bala: I probably have to apologize on both our behalf for being horrible managers at the beginning. I do not think you can really go to school for business as an undergrad because it is a very practical discipline. The only way to get management experience is by being in the fire. So, one is a word of apology, the other is a whole lot of gratitude for the people that have been with us over a long period of time as we evolved. There are people who are great for a role at a certain part of the growth curve. You have to look for fit at every step along the way. The people that have a long career horizon in the company have both leveled up and evolved over time. That is our ideal plan, where we can commit to an individual being successful in the organization with a role that evolves and grows over time. It is not always possible. It is important to know when this happens and work to figure out the next step for that person. It can create lots of issues if not handled well and proactively.

We believe in giving feedback often and as directly and as forward focused as possible. We are clear on expectations and how the employee can be successful. We have always had a flat managerial style in a sense that peer-based feedback is the most valuable for most folks. In general, 360 feedback processes are the best ways of being able to have personal reflection on what employees are doing well and having that reflection come back into a process leading to an evolution for that individual. It is important to maintain a merit-based and performance-oriented culture. Over time, if that's not the case, it becomes a rather unfair place to work in the eyes of high-performing people.

Sertl: *Is there a specific way that you describe your corporate culture?*

K. Bala: For Velan Studios, there is a core DNA around curiosity. Curiosity is one of the core attributes that we look for and hire against. There are a set of values tied to curiosity that we have telegraphed very clearly. We create breakthrough experiences that challenge convention. We are scrappy in our problem solving and are always doing it in new ways. We talk about staying curious about the world around us. Looking for new approaches to things and how the world works encourages experimentation. You have to have rigor around it. It takes mutual respect and responsibility both within the company and our customers, our players and the community. Consumers and creators are coming closer together. And creating and encouraging that community of makers and players allows us to have a tighter relationship with our community and other customers.

Velan Studios Team

Sertl: *Are there any specific listening techniques that you use with your customers to learn the pivots that are necessary in your business?*

G. Bala: There is an adage: if you ask a customer what would surprise them you will not get a good answer. The starting point is looking for insight and looking for what is new and what can be compelling. But there are different ways at different points in a product's development that you can sense if a starting point is working. As soon as we build a prototype, we put it into people's hands. And if it is not immediately obviously fun and engaging, we know it is probably not the right place to be. So, we keep grinding on it until we find that sort of single piece of magic or spark. If we feel we have something that people resonate with that can be a universal sort of fantasy, we take it forward to subsequent evolution. There are a lot of things we can do down the line but that initial spark really needs to come from inside instead of research. Eventually when the game or product ships, we heavily instrument the software, so we can figure out through usage patterns exactly what is going. The features that are being used, what is not being used, what is engaging and what is frustrating.

Sertl: *When you think of your dashboard, both tangible and intangible aspects, are there any insights in what you measure you can share for our future founders?*

G. Bala: That is a great question because so many businesses have different KPIs. Our distinction is how we measure customer delight and the value the person is getting. It is obviously important for games. In almost all situations you are trying to substitute an existing behavioral pattern for a new one. For example, take filing papers into a file cabinet and you want the behavior to become to file digitally. It takes user delight to change the daily practice. You will need even more user adoption to get this behavior to change at the mass level than your initial customers. So, an element of delight is an important measure to track and amplify. That's one way of looking at it. Another way of looking at it is asking, "who are the paying customers?" and "how are they paying?", "how much?", "how often?", and "how long do they stay with you?" So, the classic life-time value (LTV) calculation for the customer is very important. Another is, "how much does it cost you to acquire a customer?" And "what do they give you for the value that you get out of them?"

Sertl: *Was there a time when you were exhausted and wondering if it was worth it?*

K. Bala: Entrepreneurship is challenging. It certainly comes with some pretty heavy costs at times. I am not sure if I spent a lot of time thinking about whether it is worth it or not. I think it is who I am, who we are.

G. Bala: When we started Vicarious Visions, it was a very meaningful and proactive choice. After college we both walked away from lucrative opportunities in order to embrace the debt we had and go to a place where we had no customers or real professional knowledge. We were passionate about figuring it out.

Sertl: *What habits have contributed to your success?*

G. Bala: I really love to learn about things. I love to get out of my comfort zone and be in areas that I do not have much capability. It is being constantly curious about technology, entertainment, culture, science or whatever. I just love to learn about things. That has been helpful in two different dimensions. The obvious one is where good ideas come from. It also helps me to be curious about people and what they can bring to our company and our industry. It is this approach of being curious about people that I bring to our employees, our colleagues, and our board members. It also allows others to bring value to the company and contribute.

K. Bala: I have a similar answer. I am always reading about stuff every day and learning new things and trying to meet people from different walks of life. I think having a continued childlike sense of wonder is important. You might ask my wife and she will say that I'm still twelve years old. That sense of wonder is important and gets you through some difficult challenges.

Sertl: *I would appreciate you both answering this individually as well. What is the best advice you've been given?*

K. Bala: After we issued our first game, we started working on our second game. Before we got help from Mike Marvin and got angel funding, we had run a lot of debt and had to decide if we should keep going or finish college and get a "real job". I called my dad and said I was in a lot of trouble. He asked, "how much trouble?" I told him the number and there was complete silence on the other end of the phone. Finally, he broke the silence and said "I don't have that kind of money. You got yourself into this mess, get yourself out." That was that. It was important for us to take responsibility for the debt and figure our way out.

G. Bala: You have to decide who you are willing to lose money with. When you are making money, people always feel good about that. When you lose money with somebody, it starts exposing all the weaknesses. So, if you're willing to lose money with a person it means that there is something meaningful in that relationship.

Sertl: *At the last UVC event we talked about failure. If I remember correctly, you were in the process of writing a talk about failure.*

K. Bala: Yes. I did a TED Talk about failure which highlights many of our failures building the company.

Sertl: *I think a lot of great people would do more if failure was talked about more freely.*

G. Bala: It is a cultural thing. I think the way you make money influences the way you invest it. Most of the wealth in this part of the country was not made in high-growth, high-risk businesses, but within large institutions where risk taking and failure were frowned upon. It makes sense for industry sectors like health care and wealth management. In high-growth and high-risk technology businesses, we need more successes and a tighter ecosystem for evolving the culture of our region.

Sertl: *What does "pay it forward" mean to you and how are you exercising it in our region?*

K. Bala: The reason I studied at Rensselaer Polytechnic Institute was that they had the incubator center. While I was going through the interview process I was told that I needed to go to the incubator. I did not know what they were incubating, or what that even meant. When I visited, it was mind blowing to me. The idea was that people were there to help you start a business in different ways without quid pro quo or any expectation of getting something in return for it. It was mind blowing that people wanted to help, wanted you to be successful and wanted to see the community be successful. Without that guidance, not just from Mike but from many people within the community, we would not have had this success or perspective. The notion that somebody has your back while you are trying to make something work. That support, that mentoring, offering perspective, looking at the next generation of talent is key. We would not be here today without other people's investment in us.

G. Bala: I do not know if there is scientific proof of what I am about to say. It seems to me there is something quite unique about American entrepreneurship. In many places around the world, businesses are handed down from generation to generation and industries and impenetrable social networks. I've noticed here in the United States that success, especially in entrepreneurship, often has a start with a mentor. When we get an opportunity to give back, it is one of the most rewarding things that we do. Each of us allocates a quarter of our time to working with entrepreneurs. Occasionally we may invest in them as well. More than anything it is about a community contribution.

K. Bala: It is super inspiring meeting the next generation of entrepreneurs.

G. Bala: It is a way for us to get our energy.

Sertl: *Is there something important that you think we should add or a question that you wished I had asked?*

K. Bala: I grew up in Rochester. When we were in high school we started our career specialties because we met Paul Travers who was our mentor in high school. Paul is the CEO of a company called Vuzix in Rochester. We met Paul nearly 30 years ago when he was starting out with this company. He asked me, "What do you want to do?" I said, "I'd love to make a video game." Then he said, "Well, then go do it." I said, "I am just a kid and I do not know how to program." He responded, "That is no excuse. Here is a programming book." He taught me very early that you just have to remove roadblocks one by one. You really have no excuse, just go do it. That mental barrier was removed and with that we started our entrepreneurial adventure. Just to hear someone say, "You can do it," removes that mental barrier. The more we can do along those lines, to be able to drive the next generation of talent, the better off we are going to be. This region will become hugely successful when entrepreneurs, startups and new ideas come together to have a worldwide impact.

You are not always going to win.
You can always give your best.

George Chamoun

You have to be able to take risks and understand certain opportunities are only going to come around once in a lifetime.

Dan Magnuszewski

George Chamoun & Dan Magnuszewski

George Chamoun co-founded Synacor (NASDAQ: SYNC) right out of college in 1998 and spent the next two decades building it into a trusted technology and revenue partner for some of the largest video and communication providers in the world. He was an early investor in ACV Auctions and joined the company full-time as CEO a year later in 2016. George also serves as Chairman of Launch NY, a non-for-profit organization supporting the startup ecosystem for Upstate New York. He is a graduate of SUNY Buffalo and grew up outside Syracuse, NY.

Dan Magnuszewski is a co-founder of ACV auctions, a mobile auto sales platform. A Buffalo native who studied Computer Science at the University of Buffalo, Dan worked as a "technical co-founder for hire" at numerous startups in NYC and California, started CoworkBuffalo in downtown and was the first Managing Director at Z80 Labs, a technology incubator and seed fund.

Sertl: *What was your original vision for ACV Auctions and what was the genesis of the company?*

Magnuszewski: My friend, Joseph Neiman, was a used car dealer. He had a decent sized operation and was going to different auctions traveling hundreds of miles during the week to find the right inventory. There was a lot of friction in that process. People attended these physical auctions where dealers would show up, stand in a parking lot or stand in a garage and have cars come by and try to bid on them. The process was inefficient because there would be multiple lanes of cars all running at the same time. You may want to bid on all of them, but you are limited to bid only on one vehicle at a time. While I was running Z80 Labs, a seed fund incubator here in Buffalo, I had this crazy idea. I did not know if it was a good idea. I just knew I wanted to bounce it off a few people. I started first by talking to my dad who had owned a used car dealership for as long as I can remember. I understood the business. I knew about the auctions. Neiman and I just started working nights and weekends on the idea and testing the idea. We brought in our early co-founder, Jack Greco and started kind of fleshing it out and getting some validation feedback. Towards the end of 2014, we put in our notices and left our day jobs. And in January 2015, we were all in.

Sertl: *Is that the point in which you came on board, George?*

Chamoun: No. I came on initially as an angel investor in 2015. Dan, Joe and Jack were going around the community speaking to folks that would be willing to help invest in the business. I was also on the Investment Committee for Z80 Labs that Dan mentioned earlier. I had already built and taken my first company, Synacor, public. We were the first tech company in Buffalo to go public and we grew to around $100 million in revenue with hundreds of employees. Dan had approached me to invest and I came on board as the CEO in 2016. The original founders had already launched the product in Buffalo and Albany and the product market fit had already been identified. There were 15 employees and we had sold 300 cars. I came on board to help them scale the business; to roll it out across the country and grow the team.

L to R: George Chamoun and, Jack Greco, co-founder and former CFO of ACV Auctions

Sertl: *What did you learn about growing a company that is different with ACV than it was with Synacor?*

Chamoun: ACV Auctions is a marketplace business. We build out every market we go into. We had to take the lessons learned and the unit economics from the first markets we went into and build a business plan of how we would grow across the country. We first focused on seven territories and applied those lessons learned to determine the next series of markets to build. In my past life, we would launch products nationally right out of the gate because when you're launching things like TV Everywhere, or email services or content type services, you were launching it with a major company like AT&T or Verizon or HBO. You did not need to build market by market because our customers were already syndicates. What makes the ACV business model different is also what makes it fun. It is harder national markets do not necessarily replicate. Not only do we need to build a great product, a great app, but we also need to build a national team. At this point we are hiring 30 to 40 people a month. Our ACV teammates need to be fully trained because most have no experience in this industry. That balance of having to build product, capture the market and scale across the country is both challenging and fun.

Sertl: *Last year you went from 30 to 80 markets. What are you learning and how is it getting easier to scale?*

Magnuszewski: Almost every time we did something, we were learning new lessons and applying them to our next section of markets. We are mostly learning about nuances between the geographies. We are beginning to work with some of the biggest dealerships in the country. They have a totally different way of working and we are constantly trying to understand how to maximize our strategy in new markets. We are also focusing on where we are failing along the way. We build each market from the ground up with a lot of detailed attention.

Sertl: *It sounds like you "scrum the scrum" as they say in Agile and that you are incredible learners. Is there a process we could transfer to accelerate learning across our Upstate NY community?*

Chamoun: We have very detailed KPI's. We look at things like how many inspections can an inspector produce. It is not just the averages but really looking at all the functions that go into the inspection process. When we understand a process we then look at the current performance and benchmark it against the entire market potential. We are constantly diving into our numbers and keeping actual and potential results very visible. We have gotten to the point that we can truly predict what is going to happen. And we are not afraid of being wrong. You are going to be wrong part of the time. Our strength comes from our ability to forecast and learn where we missed our predictions. We have even built in a buffer in our model knowing that we will not be accurate all the time. You need to constantly predict and measure yourself against your prediction.

Magnuszewski: George says that being good listeners on the product side is as important as being good listeners on the sales side. We are not trying to jam a solution down everyone's throat.

We just try to really understand how what we do impacts others. It is not 100% perfect. We can always improve. Iteration and feedback are where we try to be smart.

Sertl: *What constraints are there on your growth plan?*

Chamoun: Hiring and training. We have training going on every single week. We are constantly trying to come up with additional ways to recruit talent across the country. We have thousands of people applying every month. Our challenge is being able to identify future teammates that match what we are looking for. And then once we do pick our teammates, making sure we train them to be able to succeed. We take it very seriously. If folks do a great job, then we know we are training them well. We hold ourselves accountable to how we can improve our training.

L to R: George Chamoun: Joe Neiman, co-founder and Chief Customer Success Officer, ACV Auctions; Dan Magnuszewski,

Sertl: *I hear you use the term "ACV teammates". Have you named your culture?*

Chamoun: We don't have the "five elements of ACV culture" on a whiteboard. However, when you walk around our Buffalo office and our offices around the country you can see people literally clapping for one another as you saw just now. Everywhere across the country, and we could show you email by email how people help one another here. There is a sense of ownership. It is bigger than the fact our employees get stock options. We are on this mission together. We see people every day, every week helping one another and learning.

L to R: George Chamoun and Jim Honsberger, Director of Inside Sales, ACV Auctions presenting to used car dealers at NIADA 2018.

Sertl: *Dan, how do you think about the engineering culture at ACV?*

Magnuszewski: From the engineering standpoint, our goal is to build world class technology and world class systems. Our goal is not to be just the best technology company in; we are trying to be the best technology company in the world.

We are approaching problems in a similar way that the best companies in Silicon Valley are. We have done a good job of bringing in expertise from Facebook, Pixar, Kroger and other companies known for their excellent systems. We know abstract ways of solving problems that can solve ten other problems later in the cycle. It is not just knowing about features. It is about how these features operate within larger systems. We spin new products and release them faster than anyone can. We have many competitors that are copying what we are doing. As we build our markets it will be harder and harder for them to keep up. We have been able to attract quality talent from some of the best companies in the world because of the way we are thinking about technology. We are also finding that many people that left Upstate New York for other careers are moving back because of what we are building and our story. As these people come back solving hard problems at scale they are even more committed to make this community stronger.

Dan Magnuszewski

Sertl: *What are some of the lessons that you learned in the funding process?*

Chamoun: There is a difference in funding between now and when I got my first startup funded. Today it does not matter where you are located. If you have experienced business leaders and if you have a great plan there is no advantage or disadvantage to where you are in the country. But you need two things. You need folks who know what they are doing, and you need a business plan that investors can really understand. When you look at our venture community we have seed resources such as Launch New York and Series A investors like Armory Square Ventures and Tribeca Ventures. As the business model solidifies and proves itself you can get Series B investors. The most important thing is to be pitching to the right investor based on the stage of the company. If there is one thing that entrepreneurs get confused about it is what these different types of investors look for and how to appropriately time the pitch. Early angel investors and seed funds allow you to raise capital without having your management team together. So long as you have an idea with some proof of concept, you can get capital. As the business matures and requires larger amounts of money, the team becomes as important as the technology or the idea. At ACV, our most recent round was a Series D and it was essential that we had our CFO in place. There is capital all over the country for Upstate New York. We need to help entrepreneurs know that there are distinctions in the funding rounds – they are not just letters.

Sertl: *George, tell us about the choice you made to go from being an angel investor to rolling up your sleeves to lead the ACV vision?*

Chamoun: I was not thinking I would ever do another startup. Startups are hard. When I joined ACV, there were 15 people and a couple hundred thousand dollars in the bank. I joined this team because I felt it was the biggest opportunity I had ever had in front of me. Andrew Shaevel is a great entrepreneur who started RSA, Remarketing Services of America. As my friend and neighbor, he articulated the ACV story very well. He said to me what ACV is doing is providing a service that sees the automotive industry in a similar way as the diamond industry. You would not buy a diamond without clarity defined. At the same time, Joe Neiman talked to me about the importance of trust transparency. The ability to build something that was not yet existing in the automotive industry and to build it on trust are the two aspects that helped me wrap my head around how big this opportunity was.

Sertl: *Given your commitment to learning, what might you have done differently now that you are on such a clear fast path?*

Magnuszewski: We ended up spending time and money with transportation companies and finance companies and many partners to enable our growth. We spent money on contingent resources to enable our technology and now look back and think we could have been more efficient with our operational capital.

Chamoun: Failure is built into our plan. I think that was probably a lesson learned from my former life. You are not always going to win. You can always give your best. We have already learned and operationalized our hindsight.

Sertl: *What are some important habits that have led to your success?*

Magnuszewski: My dad always says: "do it right the first time and you don't have to do it again". I have always tried to build things on solid foundations. I am very proud that we can attract smart people to come work for us. These technologists want to work on hard problems and they want to be using the best technologies. We have built an engineering culture of excellence. I do a ton of reading and listening to as many books as possible on Audible. I learn from the other people who have already gone through this who have the scars, who have raised hundreds of millions of dollars more in venture capital and have had successful exits. I want to really understand what they did and try to apply what makes sense to build the best in breed.

Sertl: *What has been the most useful advice you have been given along the way?*

Magnuszewski: That is a hard question. Really it is just to have a growth mindset in what you are building. Never assume that you have the right answer. If I am not the dumbest person in the room, then I have done something wrong.

Sertl: *What is your most important piece of advice for other entrepreneurs?*

Magnuszewski: You have to be able to take risks and understand certain opportunities are only going to come around once in a lifetime. You need to get past what is known and comfortable because if you are comfortable you are doing something wrong. You need self-confidence and an awareness of the next step. For me, I sold our beautiful house and started this journey with a ten-month-old child. We moved into a smaller house and it seemed like a natural thing to do. I had enough facts and knew this was a moment to capture.

To be efficient with fundraising, have a pipeline and have a process that is updated regularly.

Devin Daly

We really just set out to do one thing; we wanted to build a product that people loved, and the rest would take care of itself.

Michael Quigley

Devin Daly & Michael Quigley

Devin Daly co-founded SpinCar in 2011 with the goal of making online shopping more intuitive and engaging than ever before. With a background in venture capital consulting, Devin was afforded a unique perspective on cutting-edge software solutions. Devin earned his Bachelor's degree in Finance and Economics from Saint Joseph's University.

Michael Quigley is a co-founder of SpinCar. Prior to this venture he helped build immersive retail experiences like Destiny USA and IAC's Connected Ventures. Michael earned his B.A. in Economics from New York University.

SpinCar was a participant in the 2013 StartFast Accelerator based in Syracuse, NY.

Sertl: *How did SpinCar begin?*

Daly: For the first year and a half we were not in the automotive business whatsoever. We started experimenting in the auto space and straddled both the fashion and auto industries for about six months. It was October of 2014 when we decided to purely focus on auto. We hired Bruce O'Brien as our VP of Sales, signed up a couple of big domestic distributors and we were off to the races at that point.

Sertl: *Were you both together at the beginning?*

Quigley: Yes, that's right. The vision was simple, our thesis was that VR, or 360 imaging, would revolutionize ecommerce. Seven years ago, that sounded like a pretty wild idea; today it's increasingly obvious that will be the case. I think in essence though, we really just set out to do one thing: we wanted to build a product that people loved, and I think we thought the rest would take care of itself.

L to R: Michael Quigley and Devin Daly in SpinCar's first office in Manhattan in 2013.

Sertl: *What precipitated that clarity?*

Daly: We had hired a guy that had worked in automotive software sales and he was helping us sell into some of these fashion and gadget companies. The three of us were lamenting over beers one night after we had a sales call with a big fashion company that we thought was going to close that quarter. As we were talking we realized it was going to be another 18 months or something. And we were asking ourselves, "How can we speed up the sales cycle?"; and "What's another industry?" As we brainstormed we wondered, "Can we sell this to auto dealers?" He had worked auto software sales and so we decided to give it a shot. So we canceled our meetings for the next day with that fashion client, went on a cold call in automotive and came back with the biggest contract in the company's history. We walked into a dealer that was part of a group in northern New Jersey. We walked in, walked upstairs, presented the product without any other traction or even demos in the space and within 45 minutes we had a contract in hand.

SpinCar booth at auto dealer tradeshow.

Sertl: *What was the reason for you to continue to build in Syracuse?*

Quigley: We are both from Central New York. It is where our families and core networks are. But beyond that CNY has a lot of fabulous resources: it is so easy to tap into the local universities for growth and hiring, and in our experience in talking with entrepreneurs from larger technology markets, the willingness of the local business community and community leaders to reach out and help is actually much more heightened in nascent technology communities like CNY. We have seen many benefits from being among the first digital technology companies in CNY.

Daly: A couple of our new hires have been from high growth venture capital backed software companies and they say this by far is the best culture they have ever been a part of. I think a lot of reason for that is the duality of the New York City mindset in regard to pace and work ethic, and the family values of being in Upstate.

SpinCar office buildout in downtown Syracuse.

Sertl: *That is very cool. How are you finding the right talent for your growth plan?*

Quigley: A lot of it has been referrals. We are lucky to have been able to work with the other local companies to foster the network. We think we have a pretty special place to work. Our people end up talking about how much they love it and that drives a lot of referrals to us, which has been fabulous.

Sertl: *If you think of teaching other entrepreneurs how to be as committed to the community as you have been, what do you say?*

Quigley: It's an attitude. It's simple. It is just giving your time freely to other entrepreneurs. I take plenty of coffee meetings with people who don't even have a pitch deck yet and they just want to talk about something they are working on. Being willing to do that and not being focused on an immediate return for yourself is critical. This sort of activity works not only for yourself but also grows the network in a city.

Daly: I had a little bit of a network from growing up here and going to school here. To Mike's point, we have always tried to do right by those that we were in business with. You know the golden rule of networking, right? You give before you ask.

Sertl: *Are you finding the right skill sets in the region?*

Daly: For certain functions, definitely. When we are trying to hire Software as a Service (SaaS) executives, that is a little bit of a challenge. We are bringing people in that commute and are in some cases relocating. I think that is good for the region. We are bringing world-class executives here that have worked at blue chip SaaS companies. Our local staff are learning a lot from these types of people.

Sertl: *Can you share percentages of how many people were really from the region versus imported from the region?*

Daly: Overall, 90% of our employees are from the region. VP level or higher, 50% are from out of the region.

Sertl: *You also have a facility in New York City. How many employees are in your New York City office?*

Quigley: Engineering is in New York City. We have 25 engineers in NYC and 65 employees in our upstate office.

Sertl: *How many new hires do you expect within the next 18 months?*

Daly: As we look at our growth forecast we will add 65-70 employees in the next 18 months. We have budgeted for 19 now and know that in the next six months we will need 20 more.

Quigley: We will be almost doubling our work force.

Sertl: *If you had a wish list for talent acquisition, what would that be?*

Daly: Engineering is always tough. Finding high quality engineers. We believe that we have A+ engineers on our team and there's an adage that A+ engineers are an order of magnitude, more effective than A- or B+ engineers. So far we have been able to source those in New York City, but it is getting more challenging. Candidly, Amazon pulling out of the New York deal was good for us. We are competing with Google and Facebook to find high quality engineering talent. Our CTO does a great job and we have a good nucleus of people in place.

Sertl: *If you think of Upstate Venture Connect, is there anything structurally that we could do to make it easier for you to grow at your pace?*

Daly: We have aggressive growth plans and our ability to really deliver our numbers is in many ways contingent on our ability to hire those people. I talked to Nasir a little bit about this. We need a process to get job specifications circulated throughout our entire network.

Sertl: *What are the biggest constraints to your growth?*

Daly: Talent and capital. Talent is one we have already talked about. The way we're moving beyond it is getting creative with commuting people in, bringing people here and showing them the benefits of upstate New York and then relocating them up here. We are capital efficient which is a benefit of being in Upstate NY. Our lower operating costs and the area's low cost of living allow us to grow in profitable manner.

Sertl: *You brought up cost of living and quality of living. Is it possible to have a family and be an entrepreneur?*

Daly: Good question. I think it's easier to be single, especially in the beginning, because of the long hours and a lot of travel.

Quigley: I think our culture is such that even though people are working long hours, they are going home happy. Our culture creates people that are super engaged and pleased with their career.

Sertl: *Did either of you leave security to take on more risk?*

Daly: 100% true for both of us. I was working in a private equity consulting firm with the goal of seven figures in seven years. I went from the Ramen noodle life to earning a solid salary. It was my second year out of college when I decided to step out of that plan into what we are building now.

Quigley: Similarly, I worked for a dot-com conglomerate in the city. I had fabulous perks and worked in a building designed by Frank Gehry. It was a wonderfully fun place to work and it didn't feel like work. I studied economics and all my friends went into investment banking and were making loads of money. I was asking myself, "What am I doing?", "Am I insane?" It was a huge gamble. Both of us were asking existential questions like "Is this nuts? "What am I doing?" In the very early days it is inevitable to have thoughts like this as an entrepreneur.

Sertl: *Did you ever have the dark night of the soul?*

Daly: I think Mike and I as the leaders of the company have always been forced to be optimistic for the employees' sake and I think you get a little bit of reality distortion where you convince yourself that everything is going to work out. We simply bore down believing we would get to the other side.

Sertl: *When did you know that the two of you could not do it alone and that you needed external help?*

Daly: After we made the move in auto, we recruited a board member who had worked in the auto industry, John Miller. But that was after we had already gotten probably our first hundred customers.

Quigley: It takes a village to build a company and I think we knew that from very early on. We always knew it was more than just designing a product. It was about designing a great company and I think we knew that from day one.

Sertl: *That is exactly why we are here. What are the elements of a growth mindset?*

Daly: For what it's worth, we both played team-based sports. I always viewed our company in the same way, "We are a team and you know we are going to make this successful together and we need other players on the team."

Sertl*: Were there any unexpected situations like a Black Swan, any disruptors that caused a limit?*

Daly: I think a good point to discuss would be the massive disruption happening in the auto space. I don't know that this qualifies as a Black Swan event but there is a huge movement in the auto sector towards fully online transactions. A lot of disruption is happening. Uber and Lyft caused disruption with ride-sharing. Tesla is selling direct to consumers. There are even these companies like Carvana out there selling and allowing you to fully check out online delivering the vehicle to your door with a two-week return policy. All this disruption is causing a lot of opportunity. There's a lot of change in the way that dealerships do business and sell vehicles. This disruption has allowed us to be successful in the space.

Sertl: *How do keep your pulse on the industry with this pace of change?*

Quigley: Investing in R&D and always thinking about growth. We've transformed the company so many times already. There are transformations that the company can undergo from a business model standpoint and a market segment standpoint. Nothing we are doing is sacred.

Sertl: *How much of your P&L is R&D?*

Daly: I think it's about 35-40 percent. Something like that.

Sertl: *Who are some people that inspire you?*

Quigley: Jeff Bezos. We've loved Jeff Bezos for a long time. He went from selling books to selling everything and then to selling SaaS storage which is now half of Amazon's business. They have seemingly unrelated business lines. I think we got our "nothing sacred" type moment from them and saying like this, "There's an opportunity here. We need to be here." And they did it and it has been a massive contributor to their success.

Sertl: *Any other titans?*

Daly: I think we both genuinely appreciate Steve Jobs. I read his book probably once every 18 months or so. We appreciate the massive disruption he has caused and how involved he was from a product standpoint and the standpoint of staying relevant. I think as founders it may be easy for us to rest on our laurels, but we choose to stay active in the product and kind of where the markets are heading.

Sertl: *I think that it is remarkable that you read a book again and again. What do you get each time you re-read it?*

Daly: Yeah, just a little something here or there. Sometimes it is just motivation rereading the story. Sometimes there is a management tip. Or something simple like how he thought about design. It is not so much to get actionable tips as it is for motivation and simply enjoyment. I would just say generally part of our culture is we expect all management to read voraciously. It cascades down to the staff level with a reading program. Once a quarter we provide a book. We believe reading is a huge component of professional development.

Sertl: *What are people reading right now?*

Daly: *High Output Management* by Andrew S. Grove, *Delivering Happiness* by Zappos CEO Tony Hsieh and *The Platinum Rule* by Dr. Tony Alessandra are important books for our business culture.

Sertl: *When did you know that you wanted outside capital?*

Daly: We have been profitable for a very long time so the capital that we recently raised was all secondary capital, it was not balance sheet cash. That capital allowed existing investors and shareholders to get partial liquidity.

Sertl: *And is it easier to get funding the second round than the first round?*

Quigley: If things are going well, it gets easier with each successive round.

Daly: I think the big one is we had a maniacal focus on getting sales in the door. People stress about tweaking the product. I think what differentiated us versus dozens of other companies up after the same angel dollars was we had significant revenue coming in the door. It was like a beg, borrow, or steal type of thing. I mean we would take deals from any sector, any different type of engagement model. Our focus was trying to get that machine running.

Quigley: Yes, instead of designing for a product sale, we designed for an enterprise sale. The first two angels we brought in took nine months of getting to know each other. When an angel invests in a first-time entrepreneur, they're betting on the people more than their experience or the business model. Cultivating a tight relationship with your early investors is so important. I think this really helps close investment and it obviously makes for a much more fruitful working relationship with the investors after the close as well.

Sertl: *Who was your primary salesperson?*

Daly: Yeah. I would say we both sold. Mike has been more about expansion after the initial sale. I think that is critical. The engineering type that sits and writes code can be out of touch with the customer. They don't know what resonates. It is those little non-verbal things you need to pick up on to have a relevant value proposition. This seems to be a real pain point for most entrepreneurs. I think it is absolutely critical.

Sertl: *Do you think founders need to be sales people?*

Quigley: Yeah.

Sertl: *It sounds like customer discovery was embedded in the sales process.*

Quigley: Absolutely. We put the entire playbook together, went out hard and sold.

Sertl: *Any lessons learned in hindsight about partners?*

Daly: The best leverage to getting funding is to not need the money. I think when we realized this in our more recent round it gave us a lot more leverage. I'm trying to think about what else we learned. I think convertible notes are an interesting thing that to avoid high legal fees. It is quicker to get done and, you know, it can take some of the emotion out if it as a small company when you can push the valuation conversation down the road a little bit. If you want to be efficient with fundraising, have a pipeline and have a process that is updated regularly. Be realistic about what the probabilities are and close dates. Manage it like an enterprise sales pipeline. In terms of negotiating term sheets stuff, I don't know a whole lot. It is not over until the money's in the bank, that was a significant lesson

Quigley: If investors are not saying yes, you are getting a "no" and you need to move swiftly. I think sometimes entrepreneurs delude themselves and waste their time. VC's rarely actually verbalize "No." You need to not be afraid to move on.

Sertl: *You had shared earlier that you had two angels that took nine months to invest. How did you know it was worthwhile to nurture those relationships?*

Daly: They started spending money. When an investor begins spending money on legal due diligence it is a good sign. They started spending on what I will call accountable intelligence.

Sertl: *Are there any lead indicators for the right time to start a board?*

Quigley: Yeah. As soon as you have product market fit a board is essential and it adds another layer of accountability.

Sertl: *One of the things that I think that you do so well is the way that you created partnerships with people and some of the agreements that you made along the way. Can you tell us a little bit more about when you started building your board?*

Quigley: Yes. I am always learning and reading business books and very open to new ideas, feedback and criticism. I think our board would say that we are very coachable. Which is a very important trait to have as a first-time entrepreneur, any entrepreneur. Board structure in general will have two founders and depending on how much capital you raise one or two representing the investors and a mutually agreed independent director. In our case it was StartFast mentor John Miller who had a lot of domain expertise about the software. He had built a few successful software companies and sold them and had relationships and knowledge in the automotive software space. Some entrepreneurs might be inclined to stack the board with friendly people who will not challenge them. That would be a wasted opportunity because there are great learnings we have taken away from our board and the feedback we have been given has been invaluable to our growing the business. The most important place to be challenged is at the board level. Having a high functioning board can really act as a multiplier for enterprise value.

SpinCar offices in Icon Tower, Syracuse.

Sertl: *Are there any principles of a high- performing board that you think about that would be helpful?*

Quigley: We have been lucky that we have able to keep our board small. I think five is a great number. I think seven is starting to stretch it. It gets to be too many voices in the room. One change we made recently, in connection with this latest round of financing is that we changed the format of our board meetings. Previously it was Devin and I on the management side of the board meeting and we now take a much more collaborative approach. We bring the whole 10-person VP and C-suite management team into the meeting. I think it is important to have a connection between the people who report to you and the board. In some ways, it feels like you are giving up a bit of control to have employees have this direct connection to essentially your boss. But so much value can be added by being willing to let that sort of exchange of knowledge and ideas happen. They are emailing each other and helping each other out. I am no longer the gatekeeper. You get to a certain size and you cannot be a bottleneck anymore. You just have to accept this. Hopefully you are able to surround yourself with A players and work together as a team. I wish we had done this sooner.

Sertl: *I hear two things: transparency and decentralized leadership.*

Quigley: Yeah. Transparency must obviously come day one. Decentralization comes over time as you build a great team who are capable of being a bit more independent.

Sertl: *Can you share any techniques that you learned as you were beginning to grow that allowed you to better "let go?"*

Quigley: We put processes in place. You cannot throw up your hands and say, "Marketing is out of my purview". You have to build a process. If we are putting in a lead commitment, marketing needs to generate a specific number of qualified leads each month and meet certain requirements. Then we define what constitutes a qualified lead. If the marketing team wants to try a wacky idea to generate leads, they know their numbers and you have to trust the process. People rely on processes.

Sertl: *How did you build your culture? What mistakes did you make early on with the team?*

Daly: I would say culture is paramount for us. It has always been a top priority. It's a top priority in 2019 as well. One early big mistake was hiring people that are high performers with low company values. People can be effective in the role you have them in, but what if they don't share the same values that we have as an organization? It is tempting to want to hire a great sale closer but if he/she doesn't fit your company values, it always ends badly. We learned that lesson a couple of times.

Sertl: *How do you hire for cultural fit?*

Quigley: In the early days, everyone can sit down and interview everyone else for a single position. You get to a point where we are currently trying to codify company values. You start building the product, and then you focus on building the company that builds the product. We are at the stage now focusing on rebuilding the culture that's going to keep building the company that builds the product. It is something we are in the thick of. Our process is still manual. You cannot really beat the interview. We have questionnaires and things but have not gone so far as personality tests.

Sertl: *Sometimes A players don't work well together. How have you been able to build a team that does collaborate and share?*

Quigley: We have definitely made mistakes on the way, I think this is something we touched on earlier, having an absolute A player is not as important as having a cultural fit in our minds. A high performer alone without that fit can pollute the culture and hinder other's performance. That can lead to bigger problems. Company culture is not the construct of the daily, weekly meetings. It is the team dinners and the informal meetings where you get a sense of belonging, if things are working or not working.

Sertl: *Is there a way you describe your corporate culture?*

Daly: I think it is a familial culture. People spend a lot of time outside of the office with one another. I mean a lot of these people become best friends. I think the old school mentality was like, "It's not okay, you shouldn't be friends with people at work." We embrace and celebrate that. We have kind of a work hard, play hard type of culture. There is an expectation that you are going to do what it takes to get things done, burn the midnight oil at the end of the sales month, or to make your churn numbers. And at the same time, we like to have fun together. There is a sense of accountability. We run the business on data, and you know, that's cascaded down from the executive level down to the staff. People know their numbers and expectations are crystal clear. We definitely celebrate the victories. We ring the bell a few times a day and celebrate customer success victories. You can see on that banner the marketing team had a celebration yesterday: 200 million vehicle walk arounds!

Celebrating 200 million walkarounds. February 2019.

Sertl: *I can feel the energy. Can you tell us about your dashboard?*

Daly: We take our data super seriously. We have created alignment at the executive level as well as at the staff level that we are tracking towards three to five main KPIs. You have to be at a point to predict outcomes in the business. If something is falling behind, it is really a call for help for another manager to step in. You help that person get back on track, so you can get a green light in that particular category. Mike and I are working on a KPI for employee engagement.

Quigley: Earlier than most companies in our space, we built a robust customer success team. So, folks are proactively reaching out to customers to drive utilization with the products. They also gather feedback on what's not working with the products, insights like "where do customers see white space around the products that we could be extending the product into?"

Sertl: *What is a lesson from hindsight?*

Daly: I think one for me is go bigger, faster, right. We were always very measured in our investments. We would see something working and add a little fuel to the fire, but we never really blew it up. We are now really starting to scale this organization. The bets and hypotheses we are investing in are of a much, much larger scale, and we're able to see impact. I wish we had been a little more aggressive with some of that stuff earlier.

Quigley: It is harder to scale on the cultural fit side. I feel like I am a car, myself, accelerating as we are actually blowing through these questions.

Sertl: *How do you know when to let an idea go or quit something?*

Daly: I think we are of the mindset you never quit. I think that is why we have been successful. It was just pure grit, and determination, and persistence. There are obviously product ideas that we have that we decide to sunset. From a business perspective we are of the mindset, figure out a way to make it work, make some changes, but do not ever quit.

Quigley: I would echo that. You could argue that we quit fashion, but I think our perspective is not that we quit but reoriented the business. Anytime you quit it could just be a pivot. It is all about how you position it.

Sertl: *Are there certain things that you have done that really helped contribute to your success? Behavioral habits?*

Daly: We have off-sites with the team. We both exercise, which I think helps. Reading definitely helps.

Quigley: We have done camping trips with the team. We've done ski trips with the team. We have done boating and fishing trips.

Daly: These types of experiences help get our employees out of the office and really connect on a personal level. These deep connections are a part of our success.

Sertl: *What does "pay it forward" mean to you?*

Daly: Mike and I are both from Syracuse. We made a conscious decision to build a business here. You know, we could be somewhere warm, but we really believe in this area and providing advice to the next generation of entrepreneurs. We want to be contributing towards turning this area into a real tech economy, a software economy. We will be investing in some of the young founders of the area too.

New York City SpinCar team.

Greg Galvin

Dr. Galvin founded Kionix in 1993 to commercialize a novel micromechanical technology pioneered by researchers at Cornell University. Kionix's optical switching technology was acquired by Calient Networks and Greg continued to pursue other markets under the Kionix name. In 2009, the Japanese semiconductor firm Rohm acquired Kionix, Inc. Greg and his team have since started a new microfluidics venture, Rheonix, in Ithaca, NY.

Greg is a founding member, and former chairman, of the Finger Lakes Entrepreneurs Forum. He has a B.S. from the California Institute of Technology in Electrical Engineering and a Ph.D. in Materials Science and M.B.A. from Cornell University.

Sertl: *Take us back to Kionix, Inc. before you sold it. What was your vision for the business and why did you start the company?*

Galvin: Kionix arose out of technology that had been developed at Cornell. And at the time I was working for Cornell in a tech transfer commercialization kind of role and Cornell was trying to license the technology out and not finding anyone interested in it. Over a series of conversations between myself and the professor who's group had invented it, I decided to create a company around the Cornell patents. That technology was a manufacturing method for silicon micro mechanical devices. It goes by the acronym, MEMS. The vision, if you will, was that our process was a far better way to make MEMS devices than anyone else had at the time. At that point, commercial MEMS devices were primarily used as accelerometers for airbags in automobiles. We thought we had a better mousetrap and there was an interesting marketplace out there for the end products.

Sertl: *Did you always know that you would run a company?*

Galvin: Not at all. To me, it's always about what is interesting. People ask, you know, if I want to be X, what do I have to do to be X? They look for this very prescriptive do A, B, C, D in that order and you become X. I think it actually arises out of our educational system where all the way through from elementary school through college, there's always a, "You need to do this to get to the next step. You need to pass this grade to get to the next grade. You need to pass this exam to get into college. And you need to do this to graduate." And it's all kind of laid out for you.

I view the real world as having none of that other than, a few specialty professions. For instance, there's a very clear set of licensing requirements that I have to go through to become a medical doctor. But I find most of us tend to be opportunistic. And you say, well, this looks like an interesting opportunity at this point in time and I'm going to try that.

That was what led me to grad school in the first place. I didn't have any intention of getting a PhD. However, a professor that I had known from my undergrad days was moving to Cornell and he said, "Hey, would you be one of my first PhD students and help me get the research group started?" "Okay. It sounds like fun. Let's go do that." And when I graduated with my PhD, the norm was to go into academia or corporate research. And, to me, it was not interesting to sit in the lab and generate data and write papers for a career. I was more interested in the business side of science and technology and started looking around for opportunities that would allow me to move in that direction. Then this opportunity to start a company came up. And I said "Okay, let's go do that."

Sertl: *What is unique about being associated with Cornell and being here in Ithaca?*

Galvin: I think if you look at most tech-based startup activity, it's usually centered around a major research university. And in Central New York, we have one major research university, Cornell. First and foremost are the people coming out of it. And second are the technology and scientific innovations coming from the research.

Greg Galvin

Sertl: *What if there was a strategy to retain some of the talent? What would be the benefit for you personally and for our region?*

Galvin: The biggest benefit in the region is economic development and job creation. So, I'm sure you've noticed in your work, entrepreneurial ecosystems very much feed on themselves. And so, the more activity there is, the more activity you generate and the more success you have, the more people want to follow on that success.

Sertl: *You have 58 patents. I wonder if there is a ratio of patent per business would be an indicator of our region's success.*

Galvin: I'd be hesitant to use that as a metric just because patents play different roles in different types of businesses. If you are starting a pharmaceutical therapeutic company based on a drug compound you invented, then the patent is the company. It's absolutely creating that. That is your whole company. If you're starting a ride sharing services business, there might not be any intellectual property at all involved in it. Patents are important for certain types of businesses and largely irrelevant to others.

Sertl: *Finding the talent, leveraging the talent, how do we solve the talent problem here?*

Galvin: This company has about 82 employees at the moment. The biggest part of the talent problem here is there simply aren't a lot of jobs available here. Hence, you know, people don't tend to remain, you know those who particularly come to Cornell are lots of undergraduate or graduate students and then there is only a handful of opportunities at any given time. The opportunities may or may not match their timing and off they go. It is also unquestionably the case that Central New York is not for everyone. The person who wants to live in San Francisco or live in Manhattan, is not typically going to be attracted to Ithaca or Syracuse. So, there's a little self-selection process there. The environment we are in today is that there is brutal competition for talent everywhere. Contrary to what the government will tell you, we have vastly more job openings than we have people.

Sertl: *We often talk about startups needing capital. To create value for our readership, what are ways that you designed your company to be capital ready?*

Galvin: The answer to that lies predominantly with sales. The pitch to the investors is what the product is, what the market for that product is, and how much revenue it is going to generate. How much profit am I going to generate? And then, "Okay, Mr. Investor, here is the return you think you can get from your investment in such and such a time frame." At every stage of the company, it's all about proving out that the value proposition of the product in its market. And at the earliest stage of a company, typically, that's all PowerPoint, right? You don't yet have the product because you don't have the money to make it yet. You haven't proven out the market because you've come up with something new that there may or may not even be a market for yet because people didn't know that you could do this. The venture investment decision is really on the credibility of the people presenting the opportunity and whatever research can be done to try and validate what that opportunity is. But then it becomes questions like *do you have customers? are they buying? what does that revenue growth rate look like? Is your business scaling in the manner that you thought you were going to be able to do it?* When that is looking good, you have a more attractive investment opportunity. If that is not looking good, you will be less attractive. So, it's really all about driving that product to market and validating the proposition that you put forth in the beginning.

Sertl: *Do you have a pretty diversified portfolio of customers?*

Galvin: Rheonix has essentially four targets. The first is FDA regulated Human Clinical Diagnostics. It is the most attractive in terms of visibility and valuation, but very hard and expensive to enter because of the multi-year clinical trial process. Then there are Laboratory Developed Tests. They are also human clinical trials, but they live in an FDA loophole where they don't require regulatory approval. The downside is that we are entirely dependent on our customer doing all the work. And they control the time, schedule and the prioritization of that activity so, it is also not a very fast way to market. The third option is Applied Markets. That's the non-human markets like food and beverage testing. This has the least barriers to entry because it is unregulated. The fourth market is environmental which includes water testing. The first product we got out the door was for beer spoilage testing. But these markets are relatively smaller, more diverse, fragmented and highly cost conscious.

It turns out there's another interesting area, which is Next Generation Sequencing (NGS). This arose out of the Human Genome Project which took a billion dollars in 10 years to sequence what we can do now for a thousand dollars in hours. It turns out that there is a sample prep that precedes sticking this DNA into the sequencer. It is a time-consuming process that requires specialized laboratories and skilled personnel. We have automated library preparation for NGS on our same platform. This process was just released as a brand-new product.

Greg Galvin was named the 2014 Cornell Entrepreneur of the Year.

Sertl: *It just seems like the application for this is pervasive.*

Galvin: It is. But what I would say in reflection is it proves to be harder to dislodge the existing techniques because of inertia. Inertia is one of the most difficult constraints to our growth. When you have a startup like Uber or Lyft, they come along, and they offer up a service in this case that doesn't exist. It was a different model and it took time for the adoption to be universal. However, if it proves to be something people like and value, an idea will eventually have an exponential growth rate. We (Rheonix) can go in and replace the existing set of laboratories and skilled technicians and staff in a hospital lab. On the surface, that sounds very attractive and it is certainly much more economical. Implementing this efficiency; however, requires that organization completely change what they are used to doing. We can save a beer company fortunes in inventory costs, because we can deliver a result in five hours instead of five days. But they have been growing cell cultures in a petri dish really cheaply and doing so for the last 50 or 100 years. Our challenge is to the find the person or the organization that gets it and is prepared to champion the internal delivery model.

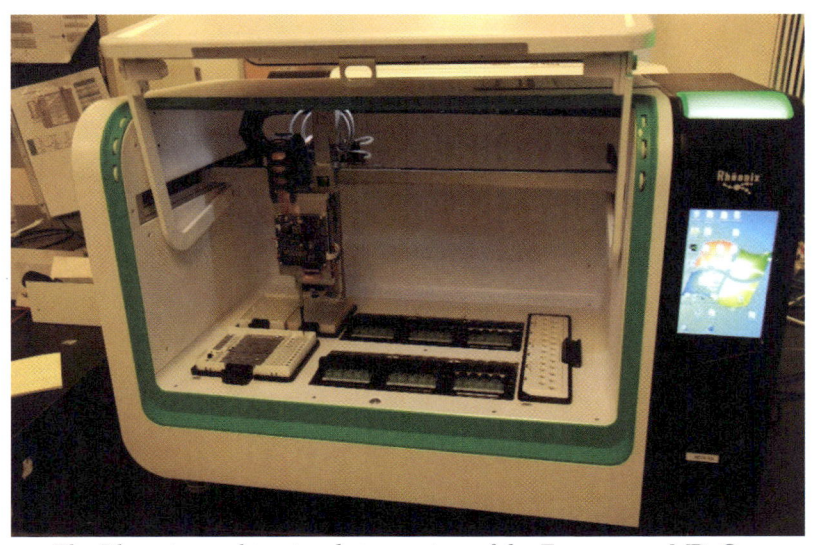

The Rheonix machine is a beta version of the Encompass MDx® workstation.

Sertl: *A lot of people who understand the value of the technology are not really good marketers. They are not able to get to the microphone or be the amplifier for you.*

Galvin: I know. There is what I've referred to in my lectures as "the inventor's myopia." You can't see past your own invention, because you created it, you think it's the greatest thing in the world, and you know how good it is. Many tech startups fall victim to the problem of not getting into the marketplace to confront reality. The market might not agree with you. You need to get out into the market sooner rather than later and find out that the customers really want a green one, not a blue one. And you also tend to find that at least in B2B type sales of technology product, customers tends to have a different opinion on what they want. And it is usually based on what burned them in the past. And so, you might have somebody whose sole focus is on the cost of your product. Whereas someone else is completely focused on your ability to deliver in quantity or somebody else's only interest is really reliability.

Sertl: *How did you train yourself and your team not to fall into "inventors myopia?"*

Galvin: I had the benefit of not being the inventor of any of the businesses I have been involved with. So, there is a level of detachment that allows me to be more receptive when the marketplace is saying, "that's not really what we want," or," that's all great, but we are not going to buy from you. You are too small."

Kionix started out making MEMS accelerometers for automotive airbag systems. Well, guess what? The auto companies are not going to buy an accelerometer from a two person start-up in Central New York to put into a safety system in their vehicles, and as a consumer driving my car, I thank God they wouldn't do that!

Kionix got its big break when we developed an accelerometer that was well suited to consumer electronics applications. And we did that fortuitously, just as the consumer electronics world started to wake up. Motion sensing ultimately ended up primarily being used by cell phones to trigger the screen rotation. So, there we were, at the forefront of supplying a product to a huge marketplace that was just beginning to adopt that product.

Sertl: *How do you prepare to get the best outcome from a sale of your company?*

Galvin: That's a difficult question. You can only sell your company when there's a willing buyer; and you don't control the buyer. Trying to sell a company is actually a pretty difficult proposition. Negotiating the sale of a company when you have an interested buyer is a different topic.

Another external force the entrepreneur does not control is what kind of buyers are out there at what times. In many cases it works out very well for the startup company because there may be buyers out there that just want to acquire your technology, your idea, and are willing to buy at reasonable numbers well before you have grown the company into a profitable revenue generating business. Those tend to go in waves or fads. For example, someone gets into the next generation sequencing business and then there four other large companies that suddenly go, "Oh crap, we don't have an NGS offering" and they snap up whatever little NGS startups they can find so that they can have that offering in their own portfolios. In that wave people are buying up NGS startups left and right. Once they have all done that, when the next NGS startup to comes along, they cannot find a buyer to save their life. For these external waves of technology acquisitions, you have to be in the right place at the right time to take advantage of them.

Another scenario is when you get more mature buyers that are looking to acquire a product line they don't have, looking to acquire a market segment they don't have, looking to acquire a customer base or a distribution network. This usually occurs when a company is more mature, generating real revenues and has a supply chain in place. The initial acquisition of Kionix was an example of the first category I mentioned. Then optical switching was hot, people needed to grab on to optical switching technologies, and there was a wave of these acquisitions and Kionix got caught in a fortuitous wave.

Later, The ROHM acquisition of Kionix was an example of the second category where ROHM wanted to get into the MEMS business. They did not have any internal MEMS product offerings. Kionix was a well-established MEMS supplier, and they could acquire it to broaden their product portfolio. And then the third category of buyer tends to be those who are buying purely for financial reasons. So typically, your company would have to be profitable with revenues in the tens of millions of dollars per year before financial buyers would have any interest. It really depends on which type of buyer and outcome you think is going to work out.

Sertl: *How did you make sure you were ready for the sale?*

Galvin: Things like your accounting practices and external audits of your financials matter. A lot of early-stage companies don't have that discipline, don't want to spend the money on it, and it's something that a buyer is absolutely going to require that would become a really big pain and potentially a deal breaker if months prior to an acquisition you had to go back through and restate all of your financial statements. Preparing properly audited financial statements all along as well as outside legal counsel ensuring all your corporate documents are correct, and you have all the requisite filings are minutiae that entrepreneurs really don't want to have to deal with. These are all things a potential acquirer will dig into very deeply. Even if you think you are getting towards an exit opportunity, setting up the so-called electronic data rooms— a cloud-based virtual repository of all the corporate documents that your company has, all your financial statements, all your legal documents, marketing materials, etcetera—is essential. This is important, so a potential acquirer can do their diligence and not actually have to physically come into your company to search through all your documents. They can do it remotely in this virtual data room. These are all very painful to create from scratch at the last minute, or easy if you keep them up to date as you go along.

Sertl: *How did you get that insight and then have the discipline follow through?*

Galvin: It really came from my board of directors. Our board of directors were pretty much all people who had been in or run Fortune 500-scale companies. That was the only world they understood. They got us to hire Wilson Sonsini Goodrich & Rosati as our first law firm which in hindsight was kind of crazy to have a two-person company with one of the world's largest technology law firms. They had us get KPMG as our external auditors and do audited financial statements every year, which again, was kind of an insane thing for a little company to be doing but it paid off in the end and we have stuck with that ever since.

Sertl: *How did you build your team and what mistakes did you make?*

Galvin: To me, hiring is an absolutely inexact science and art. How much do you know about anybody in the 30 minute or two-hour interview? We typically don't really know what the person's going to do for you for six months. And so being willing to cut somebody loose as soon as you realize that they are not working out and get someone in who will, is really important to the organization. For most people, that's a hard thing to do. The other thing that I stress to people who asked me that question is for most positions I'm worried more about the person's personality and ability to fit in the organization than their skill set. Certainly, there are some things where you need someone with a very specific skill set, that's what you're hiring them for. In early stage startups, most people wear a lot of different hats, and so it's really judging their ability to get along with the rest of the team, their belief in the mission of the organization, and the thing you're trying to accomplish, willingness to work 80 hours a week if needed and you know, hop on a plane, if needed, but whatever it is. These soft characteristics outweigh any particular skill set at the early stages.

Sertl: *Was there any demarcation of when things got more complicated from an employee perspective?*

Galvin: Absolutely. Certainly, roles and responsibilities at five people are a lot more ill-defined than at 20 or 50 or 100. The level of management hierarchy or infrastructure changes with more people. In the five-person startup there is one person in charge and everybody knows that. At 250 employees there is a management structure because one person isn't able to manage 250 employees. One of the evolutions in the organization is at what point do you have a full-time HR director? Managing employees and employee behaviors and interpersonal conflicts are not aspects I think most entrepreneurs think about when they think about starting a company. It doesn't take very many people before you'll see every form of bad behavior you can imagine.

Sertl: *Is there an employee number where you decided to make that decision for your company?*

Galvin: I would say it was probably in the 20 to 30 employee range. It becomes impossible to keep up with the HR regulations and compliance issues and so forth.

Sertl: *In hindsight, is there a decision that you might have done differently?*

Galvin: In the lessons learned category, I would say number one is raise a lot more money than you think you need, because you inevitably need more than you thought, and then probably double that again. The other one we kind of talked about earlier is to get in the market sooner rather than later. And I think that tends to be a real problem for engineering technology product company start-ups because everyone is so focused on making the widget that they are not going out and finding out is this what the marketplace really wants. It hurts because you're burning your money, which you didn't raise enough of, and then you're finding out after you burned most of your money that you don't know if the customer wants a green one or a blue one and then you have to raise more money to fund development for another six months because you didn't develop the right thing in the first place.

Sertl: *We are coming to the mindset part of the interview. Was there ever a time when you were exhausted, wondering if it was worth it?*

Galvin: Entrepreneurship is an emotional roller-coaster. One day you close a nice round of financing and you're all euphoric, and the next day you find a fatal flaw in your product, and the future looks miserable. You are perpetually running out of money. You are perpetually wondering if you are going to keep the doors open.

Sertl: *What gives you extra fuel to continue to ride those waves of up and down?*

Galvin: Initially it was the belief in the opportunity, belief in the product. Having gone through the cycle a few times successfully, I am reasonably confident and more immune to the ups and downs because I've seen them before. And even when it looks like a disaster we know we will get through this one too.

Sertl: *What is the best advice that you've been given?*

Galvin: I was told that I'm not firing soon enough. There is so much to do around the issue of personnel. From hiring to firing, to managing and motivating. Constantly staying attuned to what's working and what's not working is important. You may need to make changes either in assignments or changes in the people in the organization and doing that all sooner rather than later. Accept the hard fact that you are running a business and not every decision is going to be popular. These are not skills that people are born with and they are certainly not skills that are taught in MBA programs or undergraduate degree programs.

Sertl: *What is the best advice you have given?*

Galvin: The two we talked about earlier: raise more money than you think you need and get in front of potential customers sooner rather than later. Those are the two that I most often give out.

When we have a big problem - we coalesce like nobody's business. We have proved it to ourselves over and over and over again.

Charles Green

Charles Green

Charles Green has been a part of Rome's "cyber" initiatives at the Air Force Research Laboratory since 1997. He is a co-founder and CEO of Assured Information Security, which serves the cybersecurity needs of the Department of Defense and the US Intelligence Community. AIS has over 310 employees in 12 locations throughout the United States. Green is a strong advocate for regional initiatives and economic development. As a company, AIS and its employees volunteer to support many regional and local organizations throughout Central New York and the Mohawk Valley. Charles holds a BS degree from SUNY Institute of Technology.

Sertl: *The Assured Information Security motto is Devastating Capability. Revolutionary Advantage. Where did that come from?*

Green: We did the motto as a bit of tongue-in-cheek. We did an internal branding campaign a few years ago that was pretty informal. We were throwing around descriptions and ideas of the business and how we look at ourselves and the team grabbed specific words that they thought characterized our business and strung it together. We laughed the first time we saw the phrase and knew it was a bit pretentious. We got a kick out of it though and thought that others would get a kick out of it as well. It was funny because those who know us know that we really don't take ourselves too seriously. And those who think we are too serious about ourselves say, "You can't do that. You can't write that." And we think that is great because we got others to talk about us and that is good.

Sertl: *That is hysterical. What was your vision for the business and why did you start the company?*

Green: Oh boy, that is a heck of a story. It was 2001/2002. Four of us had gotten together and formed AIS on the side to help local businesses. We were learning so much in the cyber security domain working for the Air Force and working with people all over the world. It was kind of cool to bring this to local businesses. That was really the design and origin of the company.

A few months in, two of the founders decided to get out. The company I was working for was acquired so there was constant turnover in the management team and the direction of the company wasn't clear. The company learned of AIS and gave me an ultimatum, "You either work for us or you need to quit." I thought about it and decided I could run with it. I talked to some of my government customers and I said, "You know, if I did this would you support me?" And they said, "Yeah. We would support you." I went back to my employer and said I would like to subcontract underneath you because you treated me well and you can make some money off me. At first, they were not interested but after the management team turned over again, I ended up subcontracting under them. By this point, the third founder got out, and I moved ahead with the business on my own.

I probably could have just as easily done it by myself, be the only employee and made a good living. But the opportunities were huge. With all the turnover at the company I left, they were not that serious about cyber security at that time. Now they have national TV ads claiming that they are cyber security experts. Within months I had already picked up a couple of employees and in those first years, our growth pattern was crazy; it was like 3, 8 16, 34, 62 employees. We just kept growing at this fast rate that was insane to keep up with because cyber security was evolving so fast. It took on a life of its own.

Sertl: *How has the vision shifted over time?*

Green: We were a research and development company and that is still the heart and soul of the company. The vision has shifted as those R&D projects become more productized. The government's smallest budget line is in the R&D budget. We typically see smaller contracts under the R&D budget. As a result, we do a lot of writing to win a $100K, a $200K, or a $500K project. If you get lucky you can win a multi-year contract, one that may reach a million dollars or more. Other budget lines are much bigger. As a result, we could end up doing just as much work to win a $100K R&D contract as we do for a multi-million dollar project.

We kept seeing our finished R&D projects getting "transitioned" to big companies like Lockheed Martin, Northrop Grumman, etc. and they were getting paid to take our ideas forward. We got paid a small amount to come up with a cool idea that nobody had thought of before (R&D) but they got paid much more to maintain it once it went operational (O&M). That was not fair, so we started working really hard to try to stay with the projects as they become more productized or got integrated into bigger projects. That catalyzed us to have both an R&D team and an engineering team.

Over time, our projects started to need support in the field at the customer's on-site locations. The customers needed help desk support as well as on-site support. To meet the need, the business evolved to include more of a traditional service model where we were paid to have people available to answer questions and fix problems in real time. This created a third element of the business.

In 2013 there was a "small" thing called "sequestration" and we got hammered pretty hard. You would not know it by looking at our growth charts though. In 2013 and 2014 our reports showed steady growth, but we were positioned to grow even more than the growth charts indicated. When sequestration hit we went a full year without getting a single contract from the government. We were over-staffed and had to do one of the hardest things we have ever had to do, we needed to have layoffs. Not wanting to have to do that again, we decided to diversify and create a commercial group which is now primarily focused with our subsidiary, GreyCastle Security, headquartered out of Troy, New York.

Sertl: *Why did you choose to build in upstate New York? Was it an accident or a choice?*

Green: I would call it a little bit of both. I was born in Watertown, NY. I went to Jefferson Community College and then transplanted down here to SUNY Institute of Technology. We were not really a well-off family. Being in the SUNY system was great for me to be able to afford going to college. From there, I had a professor who put my resume in for a job working for Rome Lab. Cyber security was just beginning to be established and they were interested in finding somebody that had some knowledge in that area. I had a reputation at SUNY for let's just say "playing in the security domain."

Sertl: *I wish our readers could see your face. It looks a little devious.*

Green: It was a fun time.

Sertl: *What type of challenges do you think are unique to our region?*

Green: That's a great question. I'll be very specific to right here in Rome where we are located. There are many businesses that have spun out of Rome Lab. When BRAC (Base Realignment and Closure) happened, it was painful for the area. The traditional government contractors here were fighting over scraps. When I began, at least three businesses told me, "How dare you set up shop in our backyard". They were fighting for survival due to abrupt changes in the environment and didn't like more fish in the pond. There was a scarcity mindset.

The region is now on an upswing and newer businesses are popping up and these people feel more like I do: business is good for business. Plus, Rome Labs is growing. I think the last statistic I heard was between $1.4- $1.6 billion in revenue. Those numbers are huge and create great opportunities for us all. We have benefited greatly from that.

Upstate New York and especially Central New York and Mohawk Valley collectively together are still a bit behind the curve. When it comes to certain things such as investment capital we are not, as a region, as skilled as we need to be. I am learning myself as we have just now started to invest in other businesses.

There is not a huge support system here for people starting businesses. I kind of joke about it and say that "home" is depending on which direction I am driving today. Am I going to drive to Albany for my Vistage group meeting? Or am I going to drive to Syracuse for my meeting at CenterState CEO this week? I am getting a lot of miles on I-90. I do like the fact that CenterState CEO has consolidated a lot of functions. There are still certain things missing that could help support people and businesses in our area. I am really excited about the Thincubator and the Commercialization Academy.

Sertl: *If there was a wish list of the support you are talking about could you name the top three things?*

Green: I think regionally, if enough people could come together, it would be beneficial to have a Vistage group. EO (Entrepreneur's Organization) is another effective group that supports growth in a different way. I belong to both Vistage and EO. A group of CEOs that can share experiences with new startups and help guide them through the ups and downs of business. Bring in people with experiences from outside the area to share their thoughts and observations. It is important to have a forum to talk and share your experience on a regular basis.

Sertl: *What are the benefits of business here?*

Green: The cost of living is awesome. A house you get here for $150-200K is $1M in Maryland. I love the seasons. I go snowmobiling and a lot of folks here go snowboarding or skiing.

Sertl: *Can you tell us how many employees you have now and how many you will need?*

Green: At AIS, not counting GreyCastle Security, we're around 310 people. I believe we have planned for a total of 80 hires this year and we are halfway there already. On a positive side, a couple of new opportunities that popped up could mean another 35 additional hires, or even higher.

Sertl: *Are you finding the right talent?*

Green: We have different offices around the country now because we need to allow employees to live where they want to live. I think it is tough sometimes to convince millennials that this is a good place to live because many of them are interested in big cities and activities that are associated with that. We do pretty well when we focus on university hires and we focus on university and colleges that have similar climate zones. Our main recruiting comes from Clarkson University, Binghamton University, RPI, RIT, UMass, etc.

Sertl: *Why did you leave a secure job to take on more risk as an entrepreneur?*

Green: I was a bit of a disruptive student in high school and college and there is plenty of proof out there about that. So, when I was in the fifth and sixth grade, the vice principal used to get rid of me by sending me down to the computer room in the basement of the school. That is really where I fell in love computers, computer architecture, gaming --the whole nine yards. But, it didn't necessarily keep me from getting in trouble. The school vice principal and I got into a knock-down-drag-out argument. I told the principal that one day I was going to start my own company in computers and make $100,000 a year. The whole conversation just stuck with me and became a driver in the back of my head. When the company I was working for gave me an ultimatum, it was not a huge leap because my head was already there, and the cosmos had just presented me with this opportunity by forcing a decision. A lot of people that I was working with were extremely supportive. I had saved some personal money, got a bank loan for $10,000 and my mom loaned me some money. I was off to the races.

Sertl: *When did you realize that you couldn't do it alone and that bringing others into your vision was the best way to accelerate?*

Green: Almost immediately. I just had a computer science degree. What the hell do I know about business and then being thrown into government contracting we have this wonderful thing called Federal Acquisition Regulations which are just a massive pile of rules. I knew nothing about accounting other than Excel spreadsheets. I had to learn about overhead and G&A, direct and indirect charges, allowable/unallowable. I was getting audited by DCAA and very quickly I needed help. Dr. Popyack, one of the original founders, rejoined the company and led the research side of the business and we just kept building. Every time we ran into something we did not understand we just looked until we found someone to help us. I am no genius. It is everyone else's knowledge and expertise that made it work.

Sertl: *What have been some unexpected situations that you had to overcome?*

Green: The first problem I ran into is that a customer didn't pay the company for three months. We had two employees on the books and although I had some money saved it was not enough. Literally on the day that they finally paid me, I had run out of money. I was writing paychecks out of my personal bank account to pay the employees. The day I got paid was some sort of crazy miracle. I finally got a contracts person on board and realized that, built into the contracts, customers were required to pay us in 30 days. Then there was the sequestration situation. We had to let people go which was horrible. We never had to do that before and we lost some great people. But on the positive side, we also had employees who went to four-day weeks and helped us get through it. Every one of my senior leadership team including myself all took substantial pay cuts and literally took our salaries to pay for other people's salaries. We all came together. A year later we calculated up what money was missed and everyone was paid what they were owed. It took a while, but we made it up to everyone.

Sertl: *That is a remarkable story. How do you build resilience into your culture now?*

Green: When we have a big problem - we coalesce like nobody's business. We have proved it to ourselves over and over and over again.

Sertl: *Any acquisitions on the horizon or have you gone through an acquisition?*

Green: We have acquired three companies so far: Information Security Solutions out of Virginia, RTGX out of Maryland and then GreyCastle Security out of Troy, NY. Since then GreyCastle Security has acquired Orange Parachute and they acquired the cyber security division of EagleDream.

Sertl: *What can you teach others about the acquisition process?*

Green: In some cases surprises popped up that afterwards and we wondered why we didn't catch them earlier. It must be more than dollars and cents. You have to look at whether or not the organization is going to be a good culture fit with you and your employees. You have to take a close look at the leadership to see if they are people you want to be associated with. Make sure that if any of the acquired leadership team stay, that they truly want to be a part of what you are building. It takes more time than you think it will to integrate the two cultures. You have to wait a bit for the other team to say that they are ready to be integrated. I found that is a nicer way to do things.

Sertl: *How did you build your team and what mistakes did you make along the way?*

Green: Most of my leadership team has been with me for ten years or more and has grown up through the business. Some folks came on board with acquisitions and a few we hired into the leadership team. I have found a lot of success promoting people and encouraging people to come up through the ranks because they already understand the culture of the organization. We have a Vistage Inside program here. Vistage comes here and works with our leadership team helping them build leadership skills. There is a difference between being a leader and a manager. People can learn the management skills and manage people and direct people and even have strategy skills. But leadership is another skill and it is required to keep and motivate people. I invest heavily on leadership and it is cool because it draws us closer together and builds a real team effort. We have been doing Disney leadership training as well. What we learned in those programs is how to get different groups of people together from around the company to share stories and how to build a culture based off the idea of storytelling. Through this process, the employees built our core values and the associated definitions.

Sertl: *Then how would you describe your corporate culture?*

Green: We are a wily group of misfits and it is common for us to pull pranks on one another. One time someone's office was filled with balloons. Once I came back from a business trip during Christmas and found my car entirely wrapped with gift paper and a bow. We have fun, we work hard. Like any family, we have conflicts and we resolve them but that is all just part of life.

Sertl: *How do you listen to your customers and learn what pivots are necessary?*

Green: There are two different categories of customers for us. Traditional services and the engineering side, as well as the research side of the business. Our understanding of success criteria is exceedingly important in order for us to help our customers be successful. Their success is our success every day of the week. On the research side, we have more freedom to propose big and bold ideas that may not be quite formulated yet. We are given an opportunity to come up with these crazy ideas and see if any of them get traction. This can be a lot of fun because it allows us to be very interactive with the customer. I have always been a big fan of building and creating and imagining things, then trying to turn ideas into something real. Our R&D side of the business is very exciting.

Sertl: *How do you stay current in the market and decide what is data versus noise?*

Green: That is a great question. We have a really large group of super smart people and in some cases, we get to be trend setters and thought leaders. In other cases, we are following trends. Some of the work that we have might be new ideas incorporating blockchain or machine learning. We actively look for customers who are interested in trying something new, doing something different. In other cases, our customers come to us and for instance say, "I've got this problem and I want it done with blockchain." We do a lot of researching, go to workshops, and we are actively a part of research organizations. We know who the leaders are in the categories of technology that we are in. We have strong relationships with a lot of vendors and universities. We get access to stuff that other people do not necessarily get access to. Many times, we are asked for advice on how to accomplish solving a problem. We are very collaborative. We have relationships with fourteen different universities.

Ribbon cutting from the AIS Denver office open house in March 2019.

Sertl: *What is one decision you might have done differently?*

Green: Good question. Our first contract was in 2002 and in 17 years of doing business, it took me up until the last two years to invest in leadership training for my senior leadership team. I started investing in myself first and immediately saw a benefit. I wished I had done this enterprise wide much earlier.

Sertl: *Was there a time that you were exhausted and wondered if it was worth it?*

Green: Probably right in the middle of the sequestration. I was having some self-doubt. I am going to make a general statement that is based my observations; based on my experience with EO and with many business leaders, we are our own worst critics. I am always tearing apart every decision that I make. Asking all the time, "What did I do wrong," "What could I have done differently?" And when I grab onto a mistake, I beat myself senselessly over it. I see this is a recurring theme among every business leader that I am associated with.

Sertl: *How do you get out of that loop?*

Green: I already mentioned the time the whole leadership team came together to take pay cuts and worked a four-day work week. There were certainly critics within the crowd. Most of the employees were super supportive. It was their support that really got me through the self-doubt.

Sertl: *What habits do you think contributed most to your success?*

Green: Realizing that I'm not the smartest person in the room. I have a team surrounding me and each person brings something very important to the table. I am just a computer science guy, I am not a degreed business person. I remind everyone of that all the time.

Sertl: *What's the best advice that you have given?*

Green: The first piece of advice I would give is that if you have an idea in your head, give it a shot. What is the worst that can happen? Failure? Most "winners" do not succeed the first time. Eighty percent of businesses fail in the first five years. What have you got to lose? Give it a shot, and worst thing is that you will learn something. Best case is that you will be successful. Take a risk.

Sertl: *What does the term "pay it forward" mean to you?*

Green: I am involved with a handful of economic development projects like Mohawk Valley REDC. I am also now on the board of directors of at Mohawk Valley EDGE and Central New York Defense Alliance. There is so much activity working to support economic development in the region. I am on a couple of advisory boards for some start-up companies. I recently got involved with some of the companies that are forming out of The Tech Garden, part of the GENIUS New York Program. My big goal there is to help these companies learn from some of my ups and downs. I want to help local companies grow. No matter what our area may be lacking, working together towards the goal of growing an ecosystem that is better for all of us. Rising tide raises all ships.

The major constraints now are the same as the ones back then, namely getting the right group of founders together, identifying highly talented and motivated initial employees, and raising capital.

David Hangauer

David Hangauer

Dr. Hangauer is a cofounder of Athenex (NASDAQ: ATNX), and a leader in the field of structure-based drug design. Following doctoral studies in chemistry at the University at Buffalo, he spent ten years at Merck Research Laboratories directing drug discovery projects. In 1989, he returned to academia and joined the SUNY at Buffalo chemistry faculty. David's expertise was sought by pharmaceutical and biotechnology companies worldwide. Most notably, he led the chemistry effort at Hypnion, Inc., which was sold to Eli Lilly in 2009 for $315 million.

Sertl: *Let's talk about your early experiences in the drug industry.*

Hangauer: Before Athenex, I started two other companies. The first was Hypnion in Boston that was looking for drugs for sleep disorders. I was hired as their consultant drug designer. Our drug got to the point of getting positive phase 2 clinical trials for insomnia. Eli Lilly bought the company for 315 million dollars seven years later.

The second was Arpida out of Switzerland. That company succeeded to the point of doing an IPO on the Swiss stock exchange and raised a lot of money, over a hundred million euros. However, the phase 3 clinical trial with our lead drug didn't work out as hoped, so the stock dropped drastically, and it wound up merging with another.

Sertl: *How do drug discovery startups get funded?*

Hangauer: The whole thing started in 2000 when we started moving into the private sector. IPO's for biotech companies were rare. The model in those days for a biotech company was that you would get just enough initial angel funding to get some data. Then you would go to venture capitalists for series A and B funding. They would incubate the company to a certain point and then sell it off to a big company. That was the typical model: get started, get as far as you can on the capital you could raise, then sell the company and everyone goes their own way.

That is a common occurrence in biotech because 90% of the drugs—not just biotech, but big pharma, too—90% of drugs entering phase one clinical trials fail somewhere in the clinical trials. So, typically, you have about a 10% success rate. As an investor that is an enormous risk to invest in a 90% failure rate. So, when you do hit, you would like a big return on your investment to compensate for the risk you took. That is why drugs cost so much. The high failure rate is what leads to the very high costs of drugs that make it. The general public does not understand this because they do not see the huge cost of failures.

Sertl: *Tell us more about your unique background.*

Hangauer: I am a Buffalo native. I did my Ph.D. at the University at Buffalo in chemistry. And I went right from there to New Jersey to work for Merck. I was at Merck for almost a decade and that is where I gained my expertise in medicinal chemistry which is the chemistry of inventing drugs. After that I moved from the private sector to the academic sector to become a professor here at the University at Buffalo. In 1989, the usual process was someone would go to a university and, if they did not get tenure, they would go work for a drug company or a biotech company. It was very rare for someone to go to a drug company and then come back to academia in those days. It was an unusual career path going to the private sector first. This unusual career path provided me with a very keen awareness of how valuable it is to transfer technology from the academic sector to the private sector. And that is why I got involved in these other startup companies and I also consulted for about a half a dozen biotech and pharma companies around the world at the same time. I really had one foot in the private sector even though I was living in the academic sector with my other foot.

Sertl: *Is this a path you would advocate for others?*

Hangauer: Yes. I think you get a very keen awareness of what kind of academic research would be valuable to the private sector, and society in general. It is very grounding, and you are not living in an ivory tower. You know what can transfer out and how it could be useful. Having both private and public career experience gave me a different psychology than the typical person who goes into academics. A lot of people in academics have actually never worked a full-time job in the private sector. We did not have a culture of professors being aggressive at transferring their technology out into the private sector.

Sertl: *How did you build Kinex that later became Athenex?*

Hangauer: Buffalo was not known for biotech companies in those days. We had to do some convincing to raise angel capital in Buffalo, which we did. The key to doing that was the team we had put together as Founders. Everyone on that team had significant experience in biotech or pharma. I told you some of mine. Johnson Lau had extensive experience already having taken Ribapharm public, the second largest pharma/biotech IPO in the US at that time. Allen Barnett had been Vice President at Schering-Plough and is the person credited for bringing Claritin to the market. Interesting side note, no one thought it would work as a non-sedating antihistamine. Allen basically put his career on the line to get that drug into clinical trials. And, of course, it was Schering-Plough's biggest seller. He had a number of other commercial successes too, including Zetia. We also had Lynn Dyster who had been CEO of a local biotech company working on a diagnostic for breast cancer. The four of us together had a lot of experience in the pharma/biotech industry. We eventually convinced Buffalo angel investors that we were worth a risk because collectively we had many successes under our belt. Our successful experience was the key to getting investments from Buffalo investors. You don't raise a huge amount of money that way. You raise just enough to keep you going. We were running as a virtual company, so our overhead was very low. None of us took salaries. We rented local office space and outsourced everything in those days. Our overhead was minimal and almost everything was spent on getting results.

Sertl: *I am surprised you did not have your own laboratory.*

Hangauer: We weren't raising enough capital. It is not that we didn't try to raise venture capital money. I knew a lot of the venture capitalists in Boston and San Francisco, from all the things I did. Johnson and Allen knew a number of them as well. We did ask and all we got back was the "e" word: i.e. too *early*. Buffalo was not a place where biotech venture capitalists wanted to invest. They like to kick the tires and hence be geographically close. We did get a top-level VC to offer to fund Kinex if we relocated our company to San Diego. They would've funded us with the typical series A and B rounds, but that would have meant relocating the company. I was a full-time professor here doing this on the side. Lynn Dyster had her family here. She wasn't interested in moving. Allen Barnett's home was in New Jersey, but his wife is a Buffalo native and they had a house here. Johnson Lau was serving as Chairman of our BOD at that time and was living near Los Angeles. He was a free spirit. He could go anywhere. They offered to fund our company if we moved our company to San Diego. We said no. Later down the line we got another offer from a top tier VC in Boston who offered for us to keep the research center here in Buffalo because they knew they couldn't move me. But they wanted the headquarters in Boston. They also wanted someone they named as the CEO. Finally, they wanted to have control of the board of directors. Then my co-founders, generous as they are, said, "Dave, if you want to do it, that's your baby. Go ahead." And I said, "No. I don't want to throw my co-founders under the bus." I made the decision for personal reasons, because they were with me all this time. It turned out that it was a fantastic decision because how could I have predicted that Johnson Lau would raise so much capital later and get us to an IPO? It turned out to be much better for all of us and the company grew to about 500 employees recently spread across the globe. None of that would have happened if we had taken the venture capital.

Sertl: *Besides capital, what were some of the constraints on your growth strategy?*

Hangauer: Well, the constraints are different now than they were then. Biotech was unusual and spinning companies out of the University at Buffalo was unusual in those days. We were pioneers and pioneers don't have it easy because you're always fighting to blaze a new trail. Since then commercializing university technology has become something a lot of people want, including upper management at the University at Buffalo and the local community here. People are now trying to push and promote innovation and today there are a lot of programs, facilities, and assets to help faculty entrepreneurs. Our community is much more startup friendly now than it was back in the year 2000. I don't think people would face many of the challenges we faced back then now. Which is a good thing. Still I think the major constraints now are the same as the ones back then, namely getting the right group of founders together, identifying highly talented and motivated initial employees, and raising capital. Those are the probably the three biggest things. And I think the first one is probably the most important, it is who do you bring into this company to help drive it. Because if you don't have the right choices there, you are not going to raise the capital and attract the talent. If you do make the right choices, you have a shot at raising capital. Your co-founders and your first employees are very critical. If you make the wrong choices there, you are done.

Early Kinex Pharamceuticals Board of Directors. L to R front: Dr. Jinn Wu, Dr. Johnson Lau (Co-Founder and current CEO), Dr. Allen Barnett (Co-Founder and former CEO). L to R back: Dr. David Hsia and Mr. Chuck Lannon. Chuck Lannon was a key person for raising much of the Buffalo angel investor capital in the early and mid years.

Sertl: *What is your advice for people to get strong founders as they begin?*

Hangauer: Having people that have recognized track records and a lot of experience. People do not have to be physically located in your area. Allen Barnett's main house was in New Jersey. He did have a second house in Canada. Johnson Lau's house was in California. You need to get the best people for your company wherever they are.

Sertl: *When you began to get employees, how did you communicate the vision and get the right people?*

Hangauer: Again, it comes to connections. You can put out job openings and you'll get all kinds of people applying. You do not really know how good they are until you work with them. People can look good on paper and be a disaster when they come into the company or they may look not so good on paper but be excellent. First employees are so critical. I think it's best done through either first-hand knowledge or very trusted second-hand knowledge. You are going to have trial and error as some employees are not going to work out. I have experience working in startups and big companies like Merck. The environments are totally different. Regarding the culture in the big companies, you tend to feel like you're a small fish in a giant pond and your personal impact on the company can be very small, almost unmeasurable, just because so many things are going on. In a small company, everybody is critical, and the impact is very visible. In small companies, you tended to recruit people who were young, very energetic, and strong willed. They want to do something. They want to be given some freedom to make some decisions on their own and to contribute in the best way they think they can. You have smart, very energetic, very ambitious, people. You need to manage that energy and freedom well. If you restrain them too much, they are not going to be happy. On the other hand, if they get too much freedom, they may go off into a direction that is not beneficial to the company. The other thing you run into with small groups of very talented, strong-willed people is there are sometimes conflicts between them. They may have different visions for what to do. This means you need to be fair, open, transparent, and willing to listen to all aspects. You need to be more of a mentor than a micromanager.

Sertl: *You have about 500 employees globally and about 150 employees in Buffalo.*

Hangauer: Yes, our headquarters is in Buffalo and we have a manufacturing facility in Clarence, NY. There is a big plant being built now in Dunkirk, NY. There is a sales office in Houston and a division in Chicago that deals with generic drugs.

We have an office in Taiwan that is mostly clinical, and we have research labs in Hong Kong and collaborations with universities in Asia. The Chinese government funded the building of manufacturing plants for Athenex, similar to what New York State did for us here. We have several joint ventures too. All of this brings its own challenges because you have very different cultures to navigate in these transcontinental collaborations. There are employees that don't see each other in person as much as would be the case when there is close geographical proximity. To help address this we have a global harmonization meeting once a year. We bring in key employees from all over the world here to Buffalo. People talk about their projects and future goals. Having facilities on opposite sides of the globe also presents a unique challenge because of the time zone differences. Our CEO, Johnson Lau plays a big role in working to harmonize our globalization process. He is smart and energetic and travels the circuit visiting our facilities continuously. He is committed to sharing information, learning, and harmonizing the global operation.

Sertl: *How has New York State supported your commitment to stay in Buffalo?*

Hangauer: I started the company in Buffalo for the reasons I said earlier. Once you are successful other states and other countries start getting interested in having you move to their location. Given our commitment to Buffalo it would have been a shame to leave. However, business is business and you have to make smart business decisions. Some good offers were being made. Governor Cuomo wanted to incentivize Athenex to stay in Buffalo and utilized funds from the Buffalo Billion as a key part of that incentive. The headquarters we are sitting in is owned by New York State and leased to Athenex. We rent the whole 6th floor of the Conventus building in the biomedical corridor. New York State is also building a manufacturing plant in Dunkirk that will be leased by Athenex. The plant is a little over a two hundred-million-dollar investment. This is significant. China also wanted us and they are also building similar plants for us, but our headquarters is still in Buffalo.

Sertl: *You are bringing the world to Buffalo and Buffalo to the world.*

Hangauer: We had many top officials from China come here and tour this facility, as well as evaluate the company, before they committed themselves to working with Athenex in China. The other unique things about Johnson Lau is that he was born and raised in Hong Kong. He has been in the United States for a long time now. He was a professor in Florida for a while, he is an MD and a Ph.D., along with also being a licensed investment banker. He speaks multiple languages and he understands the culture in Asia and the US/Europe. Dr. Lau was able to connect us with the Chinese government and the Hong Kong government. Our transcontinental system that we have now is only possible because of his skill set.

Sertl: *When did you know you were ready to have an IPO?*

Hangauer: That is largely the CEO and the board of director's decision. The company was getting more and more successful, which allowed us access to capital in a different way. An IPO expands your business opportunities. Your stock also becomes a quantifiable currency of sorts that allows you to be in a position for acquisitions if that fits your business plans. Being a publicly traded company, you have a lot of regulations you have to abide by and you do not know if you are going to be successful in the IPO process. Your stock price can go up or down for reasons that don't make any sense to you. Our company had grown to the point that the board and Dr. Lau decided that an IPO was a good idea. Dr. Lau had done an IPO before. He took Ribapharm public which is the second largest biotech IPO. The largest in the United States prior to that was Genentech.

Sertl: *Are there unique KPIs that can inform the business models for emergent entrepreneurs?*

Hangauer: I can only speak about the biotech industry. What was unique is our aggressiveness of going global and connecting in Asia. I think that would probably not be something that most biotechnology companies would do. We were competing with companies like Merck and Pfizer for developing drugs to bring them to the global market. Dr. Lau shook the business model by forging collaborations around the globe and found a way to do global clinical drug development without having such a huge budget.

Three co-founders with their wives taken at the Athenex IPO on June 14, 2017.

Sertl: *How do you describe your customer base?*

Hangauer: Collaborations and joint ventures were considered our initial customer base. Medical doctors, and their patients, will ultimately be our customer base. Most of what we develop goes through prescriptions. For our generic drugs, hospitals and healthcare clinics are part of our customer base.

Sertl: *Global biotech is a pretty robust ecosystem. With that many dynamics how do you determine what is valuable information versus noise?*

Hangauer: That comes from experience. Our team has many years of experience in the pharma and biotech industry. You develop a sense of what is going to pay off. You are always titrating risk vs. value. Some of the things that might be high risk might also have a huge payoff if they work. You may want to put a certain percentage of your effort on those and then have a pretty good chunk of your effort going into things that you think are highly likely to succeed and pay off, but maybe not to the huge level. You need to balance the higher risk, higher reward, with the lower risk, lower reward research activities. It is a balancing act. It also goes back to getting the right founders and co-founders and the right upper management.

Sertl: *You have accomplished so much. Was there a time when you were exhausted?*

Hangauer: I was always exhausted because I had so many irons in the fire. I had pressure on me to produce as a faculty member. I had big teaching responsibilities with large classes of 350 undergraduates. I was trying to run a successful research program that spun off to this company. After spinning out Athenex/Kinex I completely revamped the research I was doing so there would be no conflict of interest with the company. I started a whole new research area, which is very risky for professors to do. Usually, a professor gets locked onto something and they just ride it to the end of their career if it's successful. I did a huge switch. I did not have a lot of time for writing grants. I was pulled in many directions at one time and it was tough balancing act. I was exhausted, almost all the time. I also had three children and was a single parent. Personal life and parenting have another set of big demands.

Sertl: *Wow, that's amazing. What are some habits that you formed that have contributed to your success?*

Hangauer: My management style is that of a mentor scientist. When I became a faculty member, I had tenure right away. I was not fighting for tenure like many Assistant Professors have to do in the early part of their academic careers. They can get a little hyper over that because it is a big thing. If you do not get it, professionally, you are done, at least at the level you were hoping to work at. Because I had tenure I could give students as much freedom as they could handle and allowed them to make mistakes. I do the same with employees. People learn from their mistakes far better than me telling them to do ABCD and that works, and it seems so easy, right? Let them make some mistakes and learn from them. It is not a hundred percent efficient, but it really develops students as scientists, and employees as independent productive people.

Sertl: *What is some of the best advice you were given in the early days?*

Hangauer: Back in my early days at Merck, I saw some very promising drug discovery projects get killed by upper management. The reason the projects got discontinued was more for political than scientific reasons. I saw some of the scientists persevere and continue the research with a small effort under the radar screen. In about five years or so there would be new management and they would sometimes ask if there were any really promising projects that had been discontinued under the previous management. Some of these projects would then be revived. As a medical chemist, I learned from these dedicated scientists that sometimes you need to follow your gut and instinct on what is going to succeed. Nothing in this business is easy to predict but sometimes you just get a sense that a project is likely to be successful even if you can't make an airtight case for it at the time. You just need to go with your gut. Allen Barnett did that with Claritin at Schering-Plough for example and it has worked out well for me too.

Sertl: *What is the best advice that you give to next generation medical chemists and entrepreneurs?*

Hangauer: Do not expect a biotech company to be successful in a short period of time. Our timeline was fifteen years to an IPO. At least in the biotech industry, you are typically looking at a decade-long commitment to do something significant. You need to be willing to go for the long view and do it for more than just money. We as Founders were committed for personal reasons. Our philosophy was that if we focused on drugs that benefited patients, the business would follow. I learned this lesson from Merck's CEO P. Roy Vagelos in the 1980's. Merck was the most admired company in the United States for three years in a row back then. He was always saying, "Look, put the patients first and this business will prosper."

Sertl: *What does "pay it forward" mean to you?*

Hangauer: I am currently working with a startup in Buffalo. There is a startup by University at Buffalo/Roswell Professor named Dr. Rittenhouse-Olson that I am involved with called For-Robin, Inc. Dr. Rittenhouse-Olson's sister Robin died at thirty-one years old from breast cancer. She is working to get the monoclonal antibody to clinical trials, so she has been connecting with me and Allen Barnett over the years. I have been closely involved and will serve on the board. I do not want to get paid. It is more of a choice of helping others. I do not want to do things for compensation at this point.

When I was a University at Buffalo professor I would give talks at different universities and colleges to recruit graduate students and I would always incorporate some economics into my talks.

Sertl: *What question haven't we asked?*

Hangauer: When professors in the university choose to start a company, a mistake that is often made is to think that 90% of the value in this company is their research and 10% is what other people do. I think that is a deadly error because if you do not recognize your co-founders with founder's equity to an appropriate level, you will not get the right co-founders and they won't be incentivized properly. I chose to share equity liberally when we formed Kinex. In fact, the CEO had more equity than me even though it was my technology. Initially there were four of us so that was a big dilution of founder's equity. Approaching this with generosity turned out to be a great decision. Everyone was fully motivated and worked hard to make it happen, and everybody was handsomely rewarded in the end. What should you do about sharing founder's equity? Be very generous.

I would encourage any young business to share more of the upside to get where they want to go faster. Velocity matters.

Michael Lorenz

Michael Lorenz

Michael Lorenz is a serial entrepreneur with a career spanning multiple industries and technologies. He was part of the leadership team at Ephesus Lighting which introduced LED technology into the sports and recreational venue lighting industries in 2012. Ephesus was acquired by Eaton Corp. just three years later. Now as CEO of NGU Sports Lighting, LLC Mike is partnering with Eaton to expand the LED sports lighting experience nationally across many markets. Mike holds a BS degree from Binghamton University.

Sertl: *Tell us about the beginning of Ephesus.*

Lorenz: In the fall of 2011 I was introduced to Joe Casper who was working on developing a more efficient chip for LED lighting. Joe was an engineer who had been working with solid state technology for many years and had an entrepreneurial spirit. He was interested in possibly developing a chip fab plant in NYS and was looking for someone to help him with the business startup. My partners and I were finishing the development of a new hotel complex in Syracuse and we eventually agreed to work with Joe on evaluating the market and potential business opportunity. During the evaluation process we shifted focus away from making the chip and towards the fixture side of the market. We saw the opportunity to design and build fixtures for high output applications—including sports lighting – as there was a strong need and seemingly no solutions available at that time. In 2012 we established Ephesus Lighting and from the onset until the time that we sold the company to Eaton in 2015 we maintained a singular focus – design and build the most innovative and reliable high output solid state lighting solutions for challenging environments.

Sertl: *Why did you choose to build the company in Upstate New York?*

Lorenz: Joe had family roots in Syracuse and he also appreciated what the region had to offer. I was born in Yonkers which is outside of the New York City area and I went to college in Upstate New York. I started my career working for Price Waterhouse in 1979 in Syracuse. We both knew that there were so many intangibles that Upstate New York had to offer a new business including a strong educational infrastructure and competency around technology and innovation. Given our commitment to the area it was a good place to build a new technology company.

Sertl: *Ephesus grew very rapidly. How did you design for scale?*

Lorenz: We relied heavily on the partner ecosystem that we were continuously building. We wanted to innovate fast – get to market fast – capture a leadership position fast. We considered the importance of scaling the business in all aspects of our decision-making process including supply chain, manufacturing and R&D. Joe Casper was especially adept at fast tracking innovation which gave us an early advantage in the marketplace. Additionally, we developed solution offerings that could easily be tweaked to fit a variety of needs fast tracking our ability to sell. The products we initially manufactured were designed for applications in two primary sectors: sports and challenging industrial applications. We concentrated our focus on environments that created challenges for lighting—extreme heat or other types of variables inside their buildings which would require a better designed solid-state lighting solution. For sport venues, we initially focused on indoor venues such as municipally owned arenas that housed tenants playing sports like hockey and basketball. Eventually, we evolved from sport arenas to stadiums. We quickly developed deep knowledge about the market needs and the best approach to meeting those needs. The majority of our sales were to venues that were typically owned by some type of governmental agency and operated by private companies.

Sertl: *Are there significant aspects for writing contracts for large organizations like that?*

Lorenz: That is the easier part of the story. There were many approaches that were followed to get our products into those venues – direct sales and selling under existing contracts via third parties who sold to the venues. The bigger challenge for us was centered around acquiring capital and attracting talent to a startup company designed to compete with multibillion-dollar companies in the lighting industry. We had to create a foundation around the core business that would be perceived to be credible in the marketplace. The key for us was attracting partners, both the industry-based partners and supply chain partners. We needed folks who could help us innovate and create the right outcomes and solutions in tight timeframes. We were aggressive and, I think, probably better than many startups in focusing on communicating our capability to the marketplace. We got extremely good at understanding the needs of the marketplace and the economic drivers for audiences we thought would benefit from our solutions and we were good at promoting our capabilities.

Sertl: *Can you share more about your journey for funding the business?*

Lorenz: As we began looking for investors many of our advisors told us to make a deal to maintain as much control for the founders and the people that were operating the company. We took a different approach. We gave our investors more control, more say and certainly more influence over the business. Our investors had the right to put their stock to us in a relatively short period of time if we missed our targets. We did that because we believed it would hold our management team more accountable to the performance of the company. We were highly efficient both in the capital raising process and in applying those funds to the business model. We had strong execution because we knew that if we fell short, the investor had the right to request their capital back. We also knew that if we were performing well, it would be easier to raise capital from the same investors. We raised a relatively modest amount of capital when you look at the ultimate value we created and received when we sold the company. We built a small investor group that we kept actively aware of what we were doing in the business. They were partners helping us from the growth of the company up to the time that we sold the business.

Sertl: *Can you share any numbers with us to give us perspective?*

Lorenz: Within three years we grew our revenue to $22 million and we sold the business for a significant multiple of that revenue. We raised less than $3 million in capital through investors along that journey a relatively modest amount- relative to the business growth. We were also supported by our supply chain partners and they shared in the public relations successes with us. Part of our success was being in a very high growth market. We had proven that we could be a market leader and we were extremely focused on execution of our plan. We put every major decision through our mental focus exercise to make sure it was consistent with our objectives relative to growing the company.

Sertl: *It sounds like you had a remarkable leadership team. How did you build that team?*

Lorenz: We had a great team of talented people with complementary skills. Early on we developed a core leadership team of five. Joe Casper was the innovator and responsible for developing the technology roadmap to support the market needs. I was responsible for developing the business strategy that would result in growing the company, securing funding, and creating partnerships to support the business and I handled the marketing and sales aspects of our business. Joe's wife, Amy, developed our supply chain and manufacturing program. Andrew Sussman, at that time my business partner, was critical in assisting us with corporate structure, IP and other legal issues. Chris Nolan led the engineering execution for the technology vision.

L to R: Ephesus founders Amy Casper, Mike Lorenz, Joe Casper

We also supplemented our internal team with external talent in areas such as PR, marketing and other key business areas. Many of these folks became investors and key supporters of our business. We were good at articulating where we saw the opportunity and how we were going to attack the market. We are huge believers in a focused plan. To have a viable business you need to have a clear value proposition and a clearly articulated execution strategy. Early in my career I learned a valuable lesson if you have a great vision with no plan of execution then really you just have a dream. Dreams can be fun, exciting and uplifting; however, at some point the dream needs some portion of reality and that comes with a plan.

Sertl: *I am hearing a lot about focus. How did you build focus into your culture?*

Lorenz: A lot of this came from my personal experience. Earlier in my career I was involved with a very big idea: Destiny USA. At that time, Destiny USA was the largest consumer destination concept in the country. Daily we were involved with thousands of tasks, activities and concepts. During the eight years working in that environment I became extremely disciplined around focusing on what was most important to building a company and achieving your goals. Some things we did well, some things we did poorly.

I took that whole experience with me and applied it to Ephesus. We had a team that was motivated, committed, and willing to work in lockstep around our goals. We talked about priorities often and we practiced focus as much as we could. I will give you an example. Somebody would say to us, "There are 250,000 billboards in America that need to be supplied with LED lighting and no one has been able to figure how to do that. You guys are innovative and capable around the technology. I have got a relationship with one of the major billboard owners." On face value, this sounds like it could have been a viable business opportunity. Similarly, with medical marijuana becoming popular, people approached us about building lighting that could stimulate plant growth. Again, a very viable business. We turned down both the billboard opportunity and medical marijuana. We wanted to stick to the space we were attacking and were confident we could carve out a leadership position.

Sertl: *That is unique. How many employees did you have?*

Lorenz: When we sold our first professional stadium in 2014 we had about 20 employees on staff and were supported by a large number of partners – including our contract manufacturing partner. When we sold the business in 2015 we had about 45 employees. Eventually the company grew to 60 plus employees in Syracuse who were focused primarily on the R&D, manufacturing planning and sales support aspects of the business.

Eaton Ephesus Bowling Day

Sertl: *How did you grow sales?*

Lorenz: Our sales efforts were centered in Syracuse until 2014 when we began to hire sales associates in other parts of the country. We also worked with a variety of partners who could make introductions on our behalf. We were also able to reach the market through trade shows, strategic relationships and business networking. We did all we could to get in front of the decision makers and thought leaders around the spaces that we were attacking. We also looked for applications which would be highly visible. We sought out opportunities that met those requirements. Early on we did the first outdoor LED sports lighting project in the country with Duke University. We also focused on professional venues for the NHL, NBA, NFL and MLB. Ultimately, we became the first to light many venue categories with solid state lighting including the University of Phoenix Stadium in Glendale Arizona – home of the Arizona Cardinals. That venue hosted the 2015 Super Bowl – the first ever played under LED lighting. Our strategy was to install our lighting systems in venues where people would take notice. Historical venues, prominent colleges, national championship sites and all-star games were primary targets. We worked hard to get the attention of the industry and get folks in those industries talking about us. We were interested in seeing the technology evolve and move forward.

Sertl: *Were you able to find the right talent for these capabilities?*

Lorenz: Yes. I think we did very well in the Central New York marketplace. As most people are probably aware, it is still a very significant place for folks to consider for college. There are a significant number of quality universities and colleges around Upstate New York with very good degree programs. We tapped into that resource. We worked with the NYS Center of Excellence for Environmental Systems. We worked with NYSERDA to support developing new technologies and established good relationships with many of the Upstate Universities. Amy helped drive a great summer engineering intern program which resulted in us attracting very talented people. We had high quality students that worked with us in the summer and who eventually came and worked for us full-time once they graduated. For companies that are looking for technical engineering and technology-based skills, this is a great area for resources.

Sertl: *What were some of the constraints that you overcame while growing?*

Lorenz: Our biggest constraint was the perception that lighting could only be dominated by multi-billion-dollar companies. You have companies like Cooper Lighting, General Electric and Philips and Cree, Inc. that are massive companies that have been doing lighting for hundreds of years in many cases. Trying to break into an industry as a small manufacturing company coming out of Upstate New York that could be seen as credible was our biggest challenge. Could we compete on a national scale with these major companies? We were confident that we had the greatest innovation and leadership on the design engineering side of the business. We had to overcome the stigma of being a startup competing against big companies. The larger companies were telling the market that the technology we were introducing was ten years out before it would readily available. Obviously, that was inaccurate. We set out to prove that with our technology and focus we could get there faster

Sertl: *How did you cross the perception chasm?*

Lorenz: Mostly through positioning Ephesus as a thought leader. Early on we had the opportunity to appear on some television programs. We were on Fox10 and Bloomberg News talking about our technology. We were on as many trade show panels as possible. We did as many interviews and talk shows as possible. We sought out as many situations as possible to talk about our solid-state technology and how every walk of life was impacted by lighting. High output lighting was relatively unchanged for 50 years. We asked people to question why lighting has not changed. Who was controlling the lighting? The big lighting companies. You have to step back and wonder to some degree, why it was changing so slowly. There was a lot incentive on the part of bigger companies to keep the adoption slower rather than faster and we were pushing hard against it. We knew enough about the segments of the market to try to look for applications where we could find customers who would adopt early. We got those early adopters on board and got them work with their counterparts, some of which were mid-market.

The first commercial application we sold was in the early part of 2012 and by October of 2015, three years later, we had become a market leader in a very large portion of lighting for sports applications with solid state technology.

Sertl: *Was there a time in that adoption curve where you took a leap of faith?*

Lorenz: I think that taking risks is important to grow and to develop. If you can manage risk to some degree, and I believe you can, you just have to keep your focus on what is on the other side of that risk. Being involved with a company such as Ephesus where we were developing technology that was good for the environment and good business. It was a good feeling doing something for the environment. Along that journey there were many points where we questioned whether we were going down the right path. When you are providing new lighting systems in venues that had never had that type of lighting the obvious question is, "is it going to work?" "Is it going to meet the needs of television, fans, players and operators?" In today's world of sports, almost at every level, there are multiple groups involved – all of whom have a specific and sometimes competing need. Lighting has become more important and matters a lot more than in the past. The risk of a new lighting application failing would be immediate and be very damaging to the reputation of our customers and ourselves. We were always mindful of that risk and would often watch a sporting event taking place under our lighting system – staring at the lights! We stuck to our plan and believed in what we could accomplish and fed off the energy of each other. We had an extremely passionate group of people that were motivated to prove that we could accomplish something cool and along the way enjoy the journey.

Sertl: *You talked about growing through partnerships. Can you share some things that you learned about developing effective partnerships?*

Lorenz: Partnership is often an overused term. In reality, we very seldom had contracts. We relied on the quality of the people that we interacted with to do the right thing as we would. The integrity of our businesses was based on creating a value proposition where everyone saw something in it for themselves, for their company and for their reputation. When you are able to build that kind of foundation and are willing to share the success that you are creating with the people who are helping you, it makes for a fun process. We had big companies helping us and some very small companies helping us. A lot of individuals put their reputations out there on our behalf. We had been partners at multiple levels. Probably the hardest partnership for a young company is a traditional lender or bank. In today's world, they tend have the least ability to truly help. You have to look for your support with companies that do have the flexibility and capability to step forward. I would encourage any young business to share more of the upside to get where they want to go faster. Velocity matters and we wanted to grow as fast as we could. We knew that if we could accomplish our goals, there would be significant rewards to be shared by everybody.

Ephesus Team in Phoenix lighting up Super Bowl XLIX

Sertl: *Were there any mistakes along the way about partnerships?*

Lorenz: One in particular was a co-development technology partner who offered ancillary support capability for our system. They ended up selling that company between the time we formed our relationship and when our relationship terminated. The new buyer had a different vision. It put us in an awkward position to replace that capability ourselves. It took a while to recover.

Sertl: *What can our community learn from those scars?*

Lorenz: I have always been cautious about the term *scars*. It gets in the way of you taking the risks necessary to build a company. It is often when you take that chance in life and in business that you find the best outcomes. I think it is important to understand that you are going to take risks if you create relationships with third parties. You can do your due diligence and put your best foot forward and try to make it work but there will always be the potential for failure. Pick partners that you think are going to grow with your success. It is worth the risk so long as you have thought through various scenarios and have prepared for various outcomes.

Sertl: *The value proposition to your external customers was very clear. How did you build your corporate culture?*

Lorenz: We had an internal business culture where we were working to balance between staying focused and having fun. We attempted to live that daily and with a small group it is relatively easy. It got harder as the company grew and people were living in different parts of the country and came in with different backgrounds and experiences. We did find ourselves dealing with a founder group challenge. We had eight folks with us from the beginning and a secondary group that came in to take us to the next level. We had to fight to ensure that people felt a part of the team. I think our culture was an accountability culture where we did what would could to communicate goals and hold each other responsible.

Sertl: *Were there unique KPIs for the business?*

Lorenz: None unique to us. I believe that it is important to look at a measurement of how each employee impacts the business. Sometimes the link between a person's individual performance and the overall business outcome is not clear. We worked to minimize that disconnect as much as possible and reward success and identified areas of improvement. The most important measurement for us was customer satisfaction and accomplishing things that others thought were five years out.

Sertl: *Once you proved you could lead and change the market; how did you retain your thought leadership?*

Lorenz: That is a good question. We pushed ourselves into the discussion groups and joined technical associations and the business associations where lighting was being discussed. We tried to bring our ideas about the future of lighting to the forefront. We sought speaking opportunities at events frequented by those in charge of operating venues that were influential in the types of technologies that would be adopted. When we were acquired in October of 2015, a lot of that changed. Eaton has a very large lighting business and we became part of that entity. There were many lessons we learned during that transition as a young innovative fast-moving company to become part of a larger business. In an acquisition it is easy to lose your identity. You really have to fight hard to maintain it.

Sertl: *In hindsight, is there anything you would do differently?*

Lorenz: That is a hard question for me to answer. In a word – no. I have never looked back. I always believe that you can only look one way in life. If you look back, you avoid looking forward. I prefer to look forward. We certainly learned lessons along the way and worked to apply those lessons as we pushed forward.

Sertl: *How did you decide it was the right time to sell?*

Lorenz: When we started the sports lighting business, we consciously sat down and summarized our personal, professional and financial goals. We agreed that we would take this business as far as we could, as efficiently as we could, and when we achieved our goals we would look to other options. As we got close to those goals, we began talking about possible transition options. We knew the lighting industry was dominated by very large companies and it was more important for us to see Ephesus become a global leader that could control the business long term. Our first realization of our true market potential became clear with the success at the University of Phoenix Stadium after hosting the Super Bowl. In many ways that event legitimized the company on a national scale and the press coverage gave us so much exposure. Shortly after, we started having people interested in our business. We put together five criteria that would be important in an acquiring partner relationship and we hired an investment banker to help us evaluate our options. We kept that process extremely efficient and focused because we wanted to avoid damaging the value of the company by taking too much of our time away from our execution focus. We received a few offers from some very good companies. We had to evaluate those offers against our predetermined criteria. There really is no perfect time to sell a business or pick an acquiring partner. It needs to feel right, and it needs to fit. The best time to exit a business is when you are able to see forward momentum. That is when you have the most choices.

Sertl: *What habits have contributed to your success as an entrepreneur?*

Lorenz: Well part of it is never looking back. Realizing that the taste of failure is much more bitter than the sweetness of success. Another is understanding risk – understanding the upside and downside and working to limit downside risk is critical in any venture. When you think about *risk* – it is a four-letter word with such a deep meaning at so many levels. If you want to achieve success you absolutely have to take on risk. And that risk can keep you up at night and can put you in an awkward position. It tests you at many levels. If you can deal with that pressure, you will be successful. The weakness of a startup or new idea often rests in the mindset of the founding team. Mindset tends to be the one thing that tends to get overlooked. Success is often found where others have never looked or convinced themselves not to look. I was involved – prior to Ephesus - in building a hotel in downtown Syracuse in 2010. There had not been a new hotel built in Syracuse in over 35 years. Many of my closest friends were questioning the wisdom of why we wanted to undertake that project in 2010, the worst real estate market in 50 years in the United States. We completed building a brand-new Marriott Hotel in downtown Syracuse in 2011. In hindsight, that hotel was one of the greatest successes in my career because we saw the potential, measured the risks, took the opportunity and we were able to execute. Our accomplishment was a complete team effort and paved the way for many new hotels and development projects that have occurred since that point in time. We took a step to revitalize Central New York from a hospitality perspective. It is a good example of an entrepreneurial risk - one where you could not find anyone who would take that risk - except the people on a team that who had an unshakable commitment to building that hotel.

Ephesus Team Syracuse

Sertl: *What is the best advice you have been given?*

Lorenz: The best advice that I was given is to surround myself with people that can complement my skills and to be honest about what those skills are. That advice has served me a lifetime. Another significant piece of advice was to look past obstacles that most people can never get past to see the real opportunity. Most people see only what is in front of them. Seeing around obstacles and staying mentally focused and strong was excellent advice. When you are a young company trying to figure out where you want to go it is easy to get distracted.

Sertl: *What does "pay it forward" mean to you?*

Lorenz: That is an emotional question. I have the good fortune to have had a series of people help me along my careers from the time I graduated in college to the life I am living today. I try to do whatever I can to provide that same support, mentorship, input to whoever inquires and asks me for help. Some people I help on a regular basis. I was fired from my first job and found myself in the tough situation. The result of failing and rebuilding myself required a lot of reflection and along that journey there were people that stepped up to help me. There is no way I could be this fortunate without a lot of help. Paying it forward is very important to me.

After seeing entrepreneurship up close, the bigger risk is the idea of staying in a secure position and always wondering.

Alex Zapesochny

The story is critical. You cannot marshal financing, find partners, or build a team without it.

Mikael Totterman

Alex Zapesochny & Mikael Totterman

Alex Zapesochny is the Chairman of Clerio Vision. He cofounded iCardiac Technologies and served as president and CEO until its sale in 2017. Prior to that he served as General Counsel and Director of Business Development for Lenel Systems. Lenel was acquired for $440 million by United Technologies Corporation (NYSE: UTX). Alex has an undergraduate degree from Cornell University and graduate degrees from American University and the University of Oxford.

Mikael Totterman is CEO of Clerio Vision, a revolutionary product platform for the global ophthalmic market based on technology licensed from the University of Rochester. He was a co-founder of iCardiac Technologies which sold to ERT for $230 million at a 20x return to early investors. He was also on the leadership team of VirtualScopics when it went public on the NASDAQ. Mikael has an MBA from Dartmouth and a BS degree in Industrial Engineering from Stanford.

Sertl: *You have had success with iCardiac and are now building Clerio Vision. How did it all get started?*

Totterman: Both Clerio Vision and iCardiac have been significant team efforts. Alex had just finished helping Lenel exit and I was getting tired of doing the corporate thing. We called each other around the same time ready to brainstorm on building a business together. We had wanted to work together since we were in 7th grade. We lived across the street from each other and went to middle school together and our first collaborations were around homework. So we went to University of Rochester and looked at many potential opportunities.

Mikael Totterman and Alex Zapesochny. Brighton High School science club 1988.

Zapesochny: For iCardiac there were three main co-founders on the business side including myself, Mikael, and Sasha Latypova. On the research side we had Dr. Jean Phillippe Couderc and Dr. Wojciech Zareba from University of Rochester who had already taken the technology far.

L to R: iCardiac Founders Alex Zapesochny, Sasha Latypova and Mikael Totterman. 2019.

Totterman: And they had some commercial customers which is unusual for an academic group. They had been doing work with Pfizer in addition to a couple of other large pharmaceutical companies. They had also sold a copy of their software to the US Food and Drug Administration which I think for both of us was validation that the idea was significant.

Zapesochny: A similar thing happened on the Clerio Vision side. On the business side we have myself, Mikael Totterman and Alexandra Latypova. From the research side we have Scott Catlin and Dr. Wayne Knox.

Totterman: There was a core group of five to eight individuals without whom it would have been impossible to pull it together. Some are business oriented and others are technically and clinically oriented because that is fundamental in the medical space. Healthcare commercialization is almost impossible to pull off because there is so much complexity to it. These types of companies require capital upfront and the only way you are going to be able to get that is if you have a very clear compelling story to which all of you are committed. It is the chicken and egg thing. Medical technology is so expensive that you need outside capital, to get outside capital you need a plan to deliver a large return on investment.

Sertl: *I think that is an important point. Can you distinguish between having a technology and having a compelling story?*

Zapesochny: The stories behind why things develop or how they develop are very important and become a significant part of the sales strategy. In the case of Clerio Vision, the technology had been a collaboration between the University of Rochester and Bausch and Lomb. For some key scientists at Bausch and Lomb one of their priorities was to develop this technology they saw as a next generation solution that was going to do all sorts of amazing things. However, before they could get traction, Bausch and Lomb was purchased by Valiant. Valiant's business model was to buy things up, get rid of research and development and take all extra profitability to purchase the next company. In our case it turned into an amazing opportunity where years' worth of beautifully crafted technologies were just sitting there. It was amazing to be able to step in and take over with all these years of research and ground-breaking solutions that had already been figured out. This allowed us to be years ahead.

The same thing with iCardiac, the technology had been developed 30 years ago and no one had yet taken the idea to scale. The University of Rochester decided to become an expert in this rare genetic condition called Long QT Syndrome. Although only 1 in every 7500 people have it, they decided to become the leaders of it. They had access to worldwide data and money. They created the world's largest database and for years they were doing this work on this kind of what seemed like a very small problem. All of a sudden, it turned out that the research and knowledge they had developed for this rare genetic condition was actually applicable to basically every single drug development effort out there. When the research was framed in the right story, iCardiac was able to raise money and get people excited to work with us.

Totterman: The story is critical. You cannot marshal financing, find partners or build a team without it. Building a business is like climbing uphill at a steep angle, so if there is not something inspiring that gives you the belief that you collectively can do it, it may be too hard to ever try. Fortunately, we have those stories and continue to build upon on them so that it becomes this flywheel that goes faster and faster and faster.

Sertl: *You are building Clerio here in Upstate New York and have offices on both coasts.*

Zapesochny: Yes. And in the case of iCardiac, we also have people spread out several places in the world. We have a sizeable presence in Colorado, Europe, Poland, Russia and Vietnam.

Sertl: *You could do anything anywhere and you have a track record. What makes building in Upstate New York so compelling?*

Totterman: I cannot imagine trying to do this in San Francisco for a long list of reasons. The amount of turnover that occurs with teams is just crazy. I mean people come and go all the time. There's very little employer stability. A startup requires consistency and you need to stick with it for a long period. Here you can devote time and resources to do it properly and the dollar goes further. In the Bay Area, you can no longer provide people jobs that afford them housing. In San Francisco, you are eligible for federal housing assistance if you make $125,000 a year. If that is the poverty level, how much do you need to pay team members? Not only that, how much money do you have to raise if that team does not stick around. Nobody is well off because nobody can make enough money to afford housing. Employers and employees are both better off building startups in this area. Not every kind of startup can be successfully built here, but for medical devices and research it works extremely well in this corridor. You need to recruit the right talent with a long history of patents and everything like that. Another reason is that there is more of a philosophy of getting deliverables accomplished rather than just putting in time. A lot of the Bay Area employees have an 18-hour day. You cannot focus and be effective 18 hours straight, so you know they are playing pool and ping pong and beer pong and everything else. Wouldn't you rather be at home doing something else in your free time? In my opinion, I think that is an unhealthy environment.

Zapesochny: There is a lot of happenstance as to why people build their first company wherever they happen to build it. Both Mikael and I assumed that Rochester was the place we would grow up and come visit family and friends periodically. For our careers we both moved away. Mikael spent time in San Francisco and Boston. I was in New York City and Washington DC. In each of our cases, something unexpected kind of drew us back. I got drawn back for what I thought was a one to two-year effort to help take Lenel Systems public. Mikael came back for a company called VirtualScopics.

Totterman: It seemed a perfect time to switch. As the dot-com bubble was crashing I began to think more about healthcare.

Zapesochny: Rochester is a university town. It's an intellectual town. It's a hardworking town. For the most part people do not have the egos that other parts of the country have. The people even in academia are easy to work with. A well-balanced outlook on life permeates through everything. You have to think globally.

Sertl: *In your growth phase what were some of the obstacles and how did you overcome them?*

Zapesochny: Until we got the Food and Drug Administration to sign off on this new kind of standard, nobody wanted to use the iCardiac technology. It was an uphill battle to get folks to use the technology even though it was incredibly valuable for everybody in the value stream. We had to change our thinking and look at the FDA as a partner and help them solve critical problems before that. One of their challenges was getting drugs to market in a way that was not overly burdensome to the Pharma companies.

Totterman: Most startups are about trial and error working to figure out how to position whatever it is that they are doing in a way that is compelling. In the case of iCardiac what we quickly figured out was that customers were concerned about going too far out in advance of a technology. The goal over time was to take some of our core learnings and translate them into the existing framework everybody was used to. That way using our technology would not feel like such a big leap to take and working with us was easier. We essentially figured out how to fit it into the regulatory framework.

Rochester Business Journal, March 7, 2014

Zapesochny: In some places our technology was too advanced for the market, so we had to work carefully with how we expressed and delivered benefits of the technology. We had to pull back a little and be patient. We can use cars as a good analogy. Think of the difference between assisted driving technologies versus fully automated driving. You do not go from one to the other immediately, it is too wide a leap. We had to make sure that our software was robust enough to compensate for human oversight while the technology adoption process was gradually moving forward.

Sertl: *Many talk about startup constraints being technology, capital and talent. What you are talking about is that customer adoption process was the biggest constraint.*

Zapesochny: It is hard to change the way people or other companies actually do things and to change their processes. You have to really understand their way of acting and understand their motivation in order to find an initial entry to tweaks to their process. From there you can find early adopters and build case studies.

Totterman: You cannot fall in love with new technology. You have to figure out how to package the technology in such a way that it is the easiest possible solution to slip into their existing framework. We have done a fairly good job bridging unique leading-edge technology from academia into the real world. The technologies we have leveraged were not ours. Our sole purpose is to keep tweaking these technologies until we find a use for them in society.

L to R: Mikael Totterman and Alex Zapesochny. 2012.

Sertl: *What did risk look like to you and how did you manage it?*

Totterman: It felt riskier the first time around, but again iCardiac wasn't our first startup experience. Alex had Lenel, I had VirtualScopics. We felt really confident in the team and had the ability to talk through hard stuff. We felt that if there was a way to make it happen we would figure it out somehow.

Zapesochny: After seeing entrepreneurship up close, the bigger risk is the idea of staying in a secure position and always wondering.

Totterman: I went back to corporate for a year and did not realize how different the two are. Even though the startup work is more stressful and riskier, it is yours and collectively you can make decisions, make mistakes and change course. Larger organizations for the most part are very political.

Sertl: *Can you say more about the conversations needed to build the right team to get funding?*

Totterman: I will work it backwards. We have ended up with teams that believe in information and data and believe in hashing stuff out. Individuals on our team have fairly different perspectives and enough respect for each other to hear one another out. For the most part, when people have strong convictions the conversation is about what is best for the business model. We have industry expertise on the team and we blend it with people outside of the industry, so we are not constrained by the industry's thinking about a particular thing. At my very first startup, too many people thought the same way and we would go off the rails together which felt fun until we actually went off the rails. As a result, I am much more of a believer in getting to the heart of conversations on the front of the deck.

Zapesochny: It is useful to think of difficult conversations in two categories. One is around making sure that there is real alignment and that plays into multiple aspects of building a company. The other is that people have different needs at different times.

Totterman: Alex is particularly good at the interview process, which is about psychological alignment. We have a plan with operating procedures that make us operationally aligned. We do not believe in administrative assistants and everyone does their own stuff. Others spend millions of dollars on a whole other layer while we believe those tasks should be done ourselves. The most important things are the product and customers. Just as an example, this table came from my mother's kitchen and these chairs we got for free. What I really learned from Alex as he came from Lenel to iCardiac was the importance of intellectual alignment around what it means to be a startup. Not only are we frugal and efficient, but we also do not think of ourselves in a singular role. There are no traditional departments. If we are shipping a product and we are behind on packaging or shipping, we all pivot our work to support shipping.

Sertl: *You both have experience with exits. What can you share about the sale process?*

Zapesochny: I want to reinforce the importance of alignment and the difficult discussions that happen long before an exit. As you get your key team and lead investors together, you need to make sure you are seeing things the same way. To support this, we designed two exit opportunities for iCardiac. The first was with Norwest Ventures. We designed the deal so that anybody could take all of their investment and continue, or they could sell part or sell all. It turned out beautifully because for some people that timing was critical, and they appreciated the opportunity to de-risk their portfolio.

Totterman: Quite a few folks ended up doing half and half.

Zapesochny: In retrospect that was wonderful because when the people did take money they felt a responsibility to find opportunities for other people to invest.

Totterman: What this taught me specifically is so obvious in retrospect: most startups are too focused on just top line growth.

They are not focused enough on actually making money. From Alex we learned how to run the business to be insanely profitable. For each dollar that came in, 60 cents went to cash which ended up being an incredibly attractive asset. The biggest thing I took away was the importance of building a company that not only has the capacity to exit but also has an ability to stand on its own legs at each point in time, so you are not necessarily dependent on financing.

Zapesochny: Mikael has the unique skill of timing. He knows the right investor at this moment and also who will be the best investor at a much larger state of the company. He is strong at building relationships and getting advice from some of those people, creating strong multi-year relationships. You will often hear him saying, "let me run something by you".

Sertl: *Are there any unique KPIs in your business model?*

Totterman: I would say we have the frugality KPI. We try to turn cost effectiveness into a game. We get everybody in on it as we explain that iCardiac sold for 250 million and a company with the same set of revenues would have probably sold for 24 million. There was is a 10x delta in our favor at the exit. There were other guys who sold for about 20 to 30 to 40 million who had higher revenues than we did. The only difference was our attention to top line and bottom line growth. The only difference was frugality. People presume that frugality is just not spending money. I think the larger part of frugality is how to be innovative about being much more efficient because in the end we ended up being essentially 10 times more efficient than our closest competitor.

Sertl: *What are some habits you developed early on that have helped you be successful?*

Totterman: You start with yourself. I have always had intellectual curiosity and love trying to figure where pieces of technology can be useful for society. I do not mind the digging that has to be done to truly understand how things work. I am not good enough to come up with an idea by myself and I really like working with the team to figure things out. Most of what I work on somebody has already built and validated. My curiosity has me take the technology and change the form to be something uniquely useful in the world.

Zapesochny: The one habit that served me well is the willingness and consistency about questioning things including things that seem to be already settled. I have never had a problem questioning expertise. The way I do it is to say, "I know it is obvious to you, but can you explain it to me?" Sometimes it is obvious, sometimes it is not. Context changes how things appear. It is such a useful habit because people do not want to look stupid and often do not feel comfortable speaking to a doctor, a statistician or a pharmacologist.

Totterman: This also gives the team more comfort in trying things out. We try to be very explicit in our plan that you are judged more by the fact that you did the experiment rather than if it was successful or not. If you do enough experiments and even if only 10 percent of them succeed, you are still ahead. The key is to run as many experiments as you can in a short period of time, so you have as much success as possible even if your rate of success is only 10 percent.

Sertl: *What is the best piece of advice that you were given along the way?*

Totterman: Perhaps the most important was to never stop learning. I know it sounds generic, but I have gotten that advice from various mentors, co-founders and investors. They all reinforce to never stop questioning things. That is the fun part about entrepreneurship because it is a process by which you create something that has never existed before.

Sertl: *What is one of the most important things that you hope to transfer when you are teaching or mentoring others?*

Totterman: I genuinely believe that an entrepreneurial mindset is something that can be super helpful to anyone whether they are doing entrepreneurship or not. The whole ideology of "try, evolve, learn, try, evolve, learn" is becoming more and more critical to how we compete as a society across the globe. We do not have the lowest cost of raw materials and then we do not have the best educational system across the board. But the spirit of trying and learning is a critical thing completely in our control.

Zapesochny: I think another aspect of entrepreneurship that surprises people when they enter it for the first time is the extent that you need mental toughness and the ability to handle uncertainty. There is so much unknown, yet you have to make decisions quickly and move forward.

I remember reading this book and I do not even remember the name of the author. There was a sentence or two from a leader in a marketing firm who said, "If you can, try to hire people into a position upward. Hire them for a role or title they have never had before but believe 100% they can do it. Now they have something to prove once they get on the team." I thought that was useful and have built my teams this way.

Sertl: *Are there any questions that we should have asked?*

Zapesochny: People do not seem to recognize that the entrepreneurship experience is like everything else. You want to find ways to practice in small ways all the time no matter where you are working. Unfortunately, what happens is people leave their jobs and suddenly are all in without building mental strength. There are so many ways to practice the skill of entrepreneurship. If someone younger is reading this interview, start a club. Or, start a small side business and join other entrepreneurs. Do not even worry about the financials. Most early startups do not go anywhere. These early startup experiences are so valuable. Practice however you can. Your ideas will evolve and get better over time.

Totterman: At Babson, there is this thing called Foundations of Management and Entrepreneurship. The program believes that entrepreneurship is a skill that can best be taught through experience. The class has to start a business the first year. They break the sixty-person class into groups of three. They have a week to design their pitch and four out of the twenty groups will get $3,000 to start the business. Those who do not win the pitch contest have to work in rolls to support the growth of those four companies. I think there are many opportunities to do stuff like this. And today we have Amazon and delivery models, so you do not have to do everything yourself. Start something. Entrepreneurship is a hands-on sport.

The biggest challenge for us was (that) we had to focus and become really good at one thing so that we could achieve a market leadership position.

Christine Whitman

Christine Whitman

Christine Whitman serves as the Chief Executive Officer and President of Complemar, Inc., a packaging and fulfillment company in Rochester, NY. She joined CVC Products in 1978, bought the company via a leveraged buyout in 1990, then took it public in 1999. Christine is a founder of the Rochester Angel Network and is a highly sought-after mentor for emerging companies. She also serves on the University of Rochester Medical Center Venture Advisory Board, is a trustee of the Rochester Museum and Science Center and the former chairperson of the Board of Trustees of Rochester Institute of Technology. Christine holds a BA degree from Syracuse University.

Sertl: *Please share a bit about your background.*

Whitman: I received my BA from the college of arts & sciences at Syracuse University. Having no idea what I wanted to do in my life, I initially majored in psychology and later expanded my interest to biology and other physical sciences. After graduation I took a job at the University of Rochester as a research technician in the Biochemistry Department. I thought about working towards a PhD but couldn't see myself working in a research lab for the rest of my life. An opportunity to travel across Europe gave me the travel bug. After returning, I took a job at CVC Products, a company that originated as the Eastman Kodak Vacuum Products Division. CVC was sold to Bell & Howell and then Bendix and ultimate spun out on its own through a leveraged buyout.

Sertl: *How did you evolve your impact as a product manager to be an influencer in the market?*

Whitman: CVC specialized in making products that used high vacuum technology for a variety of different applications. Vacuum distillation was a technology used by companies for making products such as vitamin E and essential oils. High vacuum technology was utilized to distill products at reduced temperatures protecting flavor and essence. My first role was product manager for vacuum distillation products and vacuum pump fluids. I had deployed a variety of separation technologies on a small scale at the University of Rochester, but product management was something I knew nothing about.

Sertl: *How did you gain your expertise in a time when access to information was not readily available and the technologies were so new?*

Whitman: I spent many hours in the evenings learning about product management, high vacuum technology and applications for the products that the company made. I also tracked down the inventors of the various technologies and built relationships with them so that I could fully understand their advantages. It was a really interesting time because this technology had been an enabler for space travel and was now becoming an enabler for the semiconductor industry that was exploding. Many of the semiconductor manufacturing tools installed today use high vacuum technology. A trip to Europe gave me a chance to see how international business worked and I had great exposure to the whole technology product sales process. I went through this cycle until I had been through all of the products in the portfolio. It was crazy in the early days of the semiconductor industry. There is something significant about being able to see these processes first hand.

Sertl: *What impacted your choice to be at the helm and buy CVC?*

Whitman*:* The demand for next generation technology coming out of Silicon Valley required lots of complex equipment to be built with short cycle times. Moore's Law was driving the pace with which chip manufacturing was evolving, and it was incredibly intense. This was driving the demand for inventing new ways to process and measure things and come up with next generation products. It was very exciting to watch all this innovation and to become more and more involved the semiconductor industry. From 1978-1990, I worked for CVC Products as an employee, starting as a product manager, director of marketing and ultimately VP of Marketing and Sales and service. During the late 1980s, the owners of CVC started the process of selling the business. As Vice President of Marketing and Sales, I was asked to host and present the business case to potential buyers. This helped me understand its value and what it would take to buy the business. I had some ideas on how to make the business better and thought, "Well, maybe I could put a team together to buy the business."

Sertl: *How did you actually make it happen?*

Whitman: First, I created an outline of what it would take to do that and over the course of the next couple of years, we were able to pull it off. I was doing my day job, taking courses in finance in the evening and travelling because most of my customers were in Silicon Valley or in Asia. And I had two young children. Good time management was key to be able to accomplish this undertaking. Fortunately, my very understanding husband was the science coordinator and a physics teacher at Brighton High School with summers off. He has been an enormous help throughout my career.

Sertl: *Did the vision shift after you bought the company?*

Whitman: It took me quite a long time to learn the business. I learned every aspect of it between 1978 and 1990. By 1988, I was becoming concerned that the company was losing ground since the owners were trying to sell the business and not interested in investing in R&D to remain competitive and grow. I made a presentation to the board basically indicating that all our customers were retiring or dying so we needed to invest in new technology. After the board meeting they came out and said, "You are now Vice President of Marketing, Sales and R&D."

Christine Whitman letter in CVC Products Catalog

Sertl: *No good deed goes unpunished.*

Whitman: Right! Right! It's the best thing they could have done for me. Shortly after receiving this additional responsibility, there was a big discovery in the area of high temperature superconductivity. One of our customers contacted us and suggested we try to get some of this material and process it with our thin film deposition equipment. It turned out that a research team at the University of Rochester had been able to make this superconducting powder. I was introduced to the team at University of Rochester's Electrical Engineering Department and we collaborated on trying to make superconducting thin films. The PhD graduate student from that program ultimately joined our company as CTO.

Sertl: *How did you fund this new capability with the superconducting powder?*

Whitman: We applied for some SBIR grants to help us secure some funding for R&D. We now had some resources to perform the R&D to develop new products. This R&D collaboration was going on while I was starting to figure out how to buy the company. My R&D collaborators from the University of Rochester also became part of the founder group that would ultimately buy the business. The new product roadmap that they helped develop was critical to our strategy for growth. It was high stress and lots of fun building the model. It is the same model I use today.

Sertl: *What were some lessons learned building that model?*

Whitman: There were many lessons learned. First is to regularly reinvent part of the business; make sure to reserve some money for R&D. Another interesting lesson was to get just the right lawyer on your team. I was referred to a great lawyer and firm that had done many deals like ours. They helped put together a good corporate structure for growth. I was also very fortunate to secure several wonderful, patient investors who believed in us. One was our Japanese distributor who was selling and installing our products for us in Japan. Another was a local investor interested in helping a female CEO. Also, we borrowed everything we possibly could, leveraged our house and fortunately my husband was willing to go along with this. In December of 1990 we were successful in acquiring CVC Products.

Sertl: *Is it easier to transition an existing company to a new model or is it better to begin from scratch?*

Whitman: We could have left CVC Products; just spun ourselves out and started a new company with new products. However, CVC had great infrastructure and a significant customer base, and we had already secured several research contracts. When you start a company, you have zero reputation. When you buy a company, customers have had both positive and negative experiences. We had to show that we were investing in new technology and could deliver next generation equipment quickly with excellent quality. When you are making a choice to buy an existing business or to do a startup, you have different issues.

Sertl: *You chose to get funding initially and then you grew it 10x.*

Whitman: Yes. In 1990, we bought an existing technology business with declining revenues and mature product lines. In order to acquire the business, in addition to securing investors, I needed to convince our bank to transfer the company's line of credit to the new company. Since I had no track record having run a company and I would be the first woman to run a technology company, I relied heavily on the credibility of my Japanese investor. He gave the bank comfort that we were worth the risk. As soon as we completed the transaction, the next big challenges hit us - the US went into a recession for the next couple years and in Rochester we had a huge ice storm that shut the operations down for one week. Also, several of our key engineers were called up for army reserve duty for the Gulf War.

Sertl: *Talk about Black Swans.*

Whitman: Lots happened all at once. That was the first year. We also had a former employee join a UK competitor with a very aggressive pricing strategy causing us to lose business that we had counted on winning. We somehow kept moving forward and continued to improve our products and do a good job. Staying focused was critical for our ability to grow the business beyond where we were. We had a wide range of products that all used high vacuum technology. We sold mass spectrometers for the analytical and chemical industries. We made distillation equipment for the food and fragrance industries and we built vacuum pumps for industrial applications and aerospace. At some point I realized we didn't have sufficient funds to be great at all these different markets. We decided we needed to sell off some of the product lines that were not our core, the ones that were not critical. Most troubling to our employees was our vacuum pump business because that's pretty much where the company started. When I sold this business, our employees thought I was insane because this was what they had been doing for the last 20 years. I had to convince them that their jobs were safe, that they would have even more interesting products to build and we would grow faster. Nobody likes to change and that caused lots of consternation but that move gave us the funding to support other products.

Sertl: *How many of your 90 employees successful made that transition?*

Whitman: We lost some. If you interview those who stayed with us, they would say, "I thought she was insane, but it worked out really well for us."

Sertl: *That is really great. In terms of doing business in New York State what challenges do you find particular to our region and how are you overcoming those challenges?*

Whitman: For CVC the biggest challenge was we had to focus and become really good at one thing so that we could achieve a market leadership position. We wanted to be one or two in our market. There were lots of other players in the semiconductor industry, so we had to keep narrowing what we were going to be really good at. We found that there were other companies that made thin film equipment in the semiconductor industry and there was a very large player that was dominant in the largest application. So, we decided to search for our unique niche. We tried six or seven different small market verticals where we would provide samples for customers interested in moving their process to production volumes. First, we sold to the R&D market, then to pilot level and finally real growth happened when the equipment was qualified for production. We worked hard to be very good at a couple of markets and that is how we were able to start growing our business. It took three to four years to get to a point where we were starting to scale, when the customers selected us for their next generation production lines. They began buying repeat orders and instead of one machine tool, they would order five.

Christine Whitman and the CVC Leadership Team 1998

Sertl: *Once you were clear you were gaining that market penetration how did you prepare for growth at scale?*

Whitman: When our business started to grow, we needed more employees. We also needed more funding because the equipment we were building was very expensive. Commercial banks were not interested in lending money to a technology company like us. To support our growth, we needed to find other sources of funding. We hired an investment bank to help secure the funding. I traveled with our Chief Technology Officer to make pitches to numerous firms and learned the valuable lesson that it didn't matter whether I was selling the most advanced technology solution or toilet paper as long as we got the business model right. Our investment banker helped us figure out how to fine-tune our story. For that round of funding in 1994, we did a strategic investment from our biggest customer raising about nine million dollars to get us to the next level. By 1997 we were ready for another round of funding as we were growing more quickly. We started to get ready for an IPO in 1997.

Sertl: *What can you teach us about getting prepared for an IPO?*

Whitman: With an IPO, it is black and white. The window is either open or shut. We missed the market window for that economic time period. So, we had to put our IPO "on the shelf" and went out for a private equity raise. We raised another ten million dollars in 1997.

During this growth period, we needed two things, money and people. So, when you ask about regional challenges, at that time our state, county and local city governments were squabbling with each other. There was very little focus on helping businesses. You could not get much help from anybody in the public sector. Fortunately for me, RIT was just down the road and RIT had made a great investment in their microelectronics facility in partnership with Texas Instruments. And they had built a beautiful lab that they used to train microelectronics engineers.

Sertl: *How did you leverage Rochester in your strategy?*

Whitman: We were able to hire many employees from RIT. Their only problem was that they had no experience. Trying to recruit people from the semiconductor industry to Rochester was almost impossible. Ultimately, I had to open an operation in Fremont, California where we set up a local applications lab and an R&D facility. We also set up a rapid prototyping facility outside of Dallas, Texas. Our Rochester operation was where we did the manufacturing and we needed all types of employees. One great asset that Rochester offered was high quality machine shops. We needed a huge amount of high precision machining to be done and it was great to tap into our local market. Ultimately, we figured out our own method of promoting Rochester to try to convince people to come. When we were successful recruiting employees they loved it and I could never get them to go back if I wanted to transfer them back to the West Coast. Rochester is a great place to raise a family. It is convenient, and the schools are good. It is a magic little place in that regard because we have many of the amenities of a big city but none of the hassles. And then we have all these universities that we can pull employees from. I was pulling from RIT and the University of Rochester. St John Fisher has great business people too, so I was pulling from there as well.

Sertl: *While you were growing how many people were you hiring?*

Whitman: When we were doubling our revenue each year, we had to hire about a hundred people annually. Well, to bring in one hundred people, we had to interview three hundred people. Those three hundred people needed to be interviewed by three employees, so we were doing nine hundred interviews in that year. And then once they were hired, we had to train them while we were really busy just to trying to get our product out the door. It was important to have documented processes in place so that when we brought these new people in, they could follow the same procedures that our other employees had followed.

Sertl: *As you were growing you also started acquiring other companies?*

Whitman: The first company we acquired was Commonwealth Scientific, located in Alexandria, Virginia. They made equipment that employed a different mechanism for depositing thin films and some of our customers were looking for equipment that could use both technologies in one multi-chamber machine always maintaining high vacuum. We wanted to integrate their technology into our equipment. I was fortunate to have a member of our Board of Directors who had extensive experience with acquisitions help me close this deal.

Sertl: *Right. Many of the people that we hope read this will be building to be acquired or building to acquire. Are there some best practices that you learned that make these types of transitions easier?*

Whitman: Yes, absolutely. I think it's very important to put a transition team together when you acquire another operation. We prefer to call it a merger and it is important to respect that you are putting two proud cultures together. The transition team needs to have people from both companies on it and some decisions should be made jointly. One basic challenge with an acquisition is that it is usually justified based upon finding "synergy". Some of the jobs in the combined business can be handled by one of the two companies; for example; one finance department or one HR department. This means that the combined entity saves money by eliminating the duplicative jobs. This gets tricky. What products are duplicated? Which product do you select long-term? It's a challenging process to go through. We weren't perfect about it. We tried as best we could to be thoughtful about it.

Sertl: *How many employees were there at the time of the IPO?*

Whitman: We had about six hundred employees in four locations. We were in Alexandria, VA; Garland, TX; Fremont, CA; and Rochester, NY.

Sertl: *Can you share any learning that you had on how to manage multiple locations?*

Whitman: I traveled to other locations. I kept an apartment in Alexandria and apartment in Fremont. So, I was on the road all the time.

Sertl: *How did you manage your resilience during all of these transitions? Is there anything we can all learn about the mindset to conquer so many different challenges?*

Whitman: It does take unbelievable perseverance. Lots of bad things happened along the way or things happened that I didn't expect. For me, I just try to look the data and figure out how we can optimize from wherever we are. I try to take emotion out of it as much as possible. I try not to waste time looking back or blaming anybody or holding grudges—no room for all that negative energy. Rather my approach is to say, "Okay, this is the situation we are in today. How do we move it forward?" Be prepared to do some pivots. If you can't make one thing work, have a couple of back up positions.

Sertl: *Right. Am I allowed to ask how many hours did you get to sleep at night? I mean how you manage your complex schedule?*

Whitman: I did not sleep a lot when I was younger because I was switching times zones all the time. I can sleep anywhere, and I go to sleep instantly. I attributed it to being a mom because I had to get up every two hours to feed the baby. I had to learn how to sleep whenever I could. In my early years I was too much of a control freak. I had to learn to let go and I had to learn to delegate. Some things were just good enough and I learned that other people's ideas would get the same result. Listening is a skill I had to learn. I brought in some coaching help along the way and did a 360 review to gain perspective on being more effective as a leader. We would do offsite meetings to try to figure out what each of our team members was good at and I learned to give my team members work they liked doing rather than force people into traditional roles.

Christine Whitman in Japan for CVC 1988.

Sertl: *How do you have your team listen to the market and with your team being global how do you handle contextual intelligence?*

Whitman: In the case of CVC, I served on trade industry and technical boards. I served on the Board of Directors of SEMI/SEMATECH. That was a trade organization where I would come together with my competitors and customers and we would talk about big industry challenges. I was fortunate in some respects because there was a shortage of women to be recruited to serve on these terrific boards. I try to stay as connected as possible and then try to get my team members to serve on committees of these boards. This allows the team to grow and also allows them to stay really connected to the industry. When you serve on these technical boards, frequently, you are sitting next to your customers. It's solving a big problem.

Christine Whitman and the CVC Board of Advisors 1996

Sertl: *I am going to shift now and talk about you as a mentor. Is there anything in hindsight you might have done differently?*

Whitman: I wish I had listened more, sooner. I would also say delegating sooner would have been beneficial. It took me a while to build that rhythm that worked. I did waste time early on getting the wrong advisers. Having the right banker and honing in on the right legal assistance. Hiring too. It is hard to hire the best people. I love all the people I've hired, but sometimes over time, their interests change so sometimes I didn't move people out of a role fast enough.

Sertl: *Are there any particular habits that have contributed to your success?*

Whitman: I am a very goal-oriented person. For my entire career, I have always developed a strategic plan for all my businesses and anything I do. We have some big goals and we develop strategies and each member of the team signs up for very specific goals. We do this every year, we keep our plan updated and we look at ourselves on a quarterly basis.

Sertl: *I like reading this on your Complemar plan: Peace of mind, delivered.*

Whitman: Exactly. It all starts with a vision and a mission and our values. I tell all new employees to refer to the core values of our business. And that's what guides us to move forward and that guides employees within the organization to make decisions independently without having to check in. They know that they're making the right decisions as long as they are following the core values of the business. I have followed this approach for all organizations that I have led. It just helps me manage businesses.

Sertl: *What is the best advice that you had been given?*

Whitman: The best advice I was given was to determine my values and set personal goals; then live every day based upon those goals and values. In 1993, I took a trip to Whistler in Vancouver, Canada for a technology conference with a bunch of really successful tech business leaders. There was a panel discussion where they were talking about what they had done right, what they had done wrong. Most of them regretted not setting their values and personal goals early on; they were completely absorbed in building their businesses until their businesses started getting out of control. They described the negative impact on their marriages and their health. These were extremely successful serial entrepreneurs, company names that we all know, and each of them said in their own way, that they had to set values and goals to get themselves grounded and to a level where each could manage their worlds. That is when I sat down and did the same. It has served me well ever since. That was the best advice I ever got.

Sertl: *What it the best advice that you have given?*

Whitman: The idea of setting personal goals and following a set of values. You have to walk the talk. Another important point I give is the importance of perseverance. Just like the nursery story "Little Red Hen." You are all on your own in the beginning, planting the seeds, watering, cutting and grinding the wheat. Many around you are happy to join in to eat the bread once it has been baked or help you once you have succeeded. So just keep moving forward, believe in your vision and what you are trying to accomplish.

Sertl: *There is so much here. Is there a question that I should be asking that I haven't thought to ask you?*

Whitman: I would just like to say that entrepreneurship is not for everyone. I just described my life and it is not about balance. There is no balance if you are running a growth company. You can find some balance in slower growing, small businesses. When you are running one of these fast growth initiatives, your investors, your stakeholders, your employees, your customers don't care if you aren't feeling well today or how much time you spend with your family. You have to make sure payroll is covered, that your customers love you, that your shareholders are seeing their path to return on investment. If you are not prepared to take this responsibility, then you should not do it because you will be miserable. You can be a member of the team but don't be the CEO because you do have to own it and make sure you get it done. Time management, as I said is the key. Just let go of the stuff that is not mission critical. You have to have a passion for growing things and trying to make things better. It should not be about the money. Money comes once these things are in motion.

About the Editors

Jennifer Sertl

Business strategist Jennifer Sertl is an internationally recognized influencer in social media and thought leader in the emerging field of corporate consciousness. She is president and founder of Agility3R, a strategy company dedicated to strengthening enterprise organizational capital. She is co-author of *Strategy Leadership and the Soul* published by Triarchy Press. As a former best practice chair for Vistage, she helped 23 Rochester based companies design for growth. She is part of Swift Banking's Think Tank called *Innotribe* and community curator for The International Center for Information Ethics.

Jennifer received her undergraduate degree in English and Philosophy from University of Colorado, Boulder.

Nasir Ali

Nasir has been mentoring and investing in Upstate NY founders for 15 years. He launched The Tech Garden incubator in 2004, followed in 2007 by the Seed Capital Fund of CNY, Upstate NY's first angel investor fund. In 2010, he joined TriNet founder Martin Babinec to form Upstate Venture Connect, a 501c3 non-profit with a mission to connect high growth founders with the resources needed for success.

Nasir is a co-founder and Managing Director of StartFast Venture Accelerator, Upstate NY's only private capital-backed startup accelerator program. Ali's investment portfolio has raised well over $140MM since 2007. He is advisor to numerous entities including the NYS Innovation Venture Capital Fund, Next Gen Venture Partners, Fitzgate Ventures and a frequent speaker on funding challenges for startups.

Nasir received his undergraduate degree in Physics from Princeton University and an MBA from Yale University.